D0206975

WHO
KNEW?

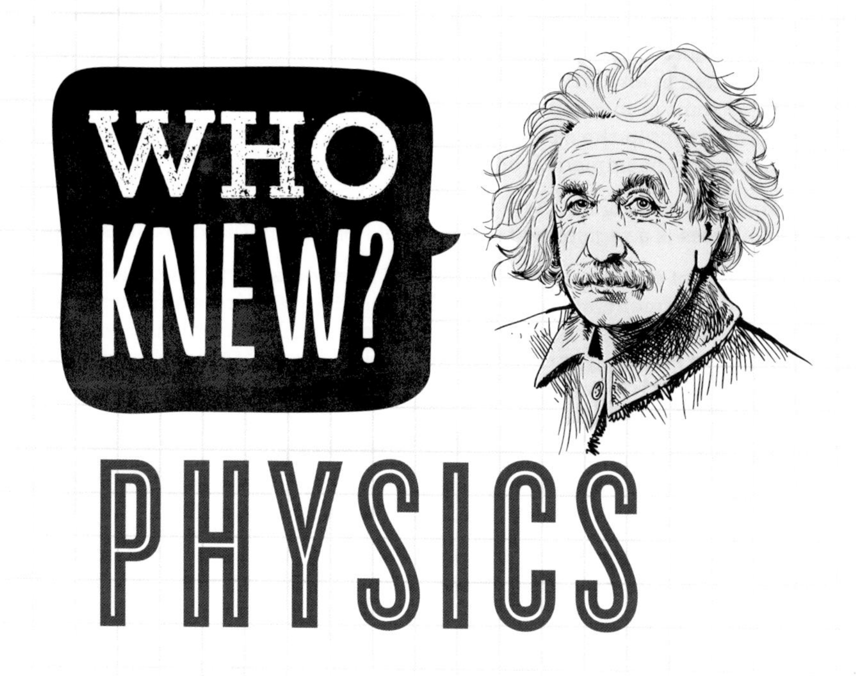

James Lees

CONTENTS

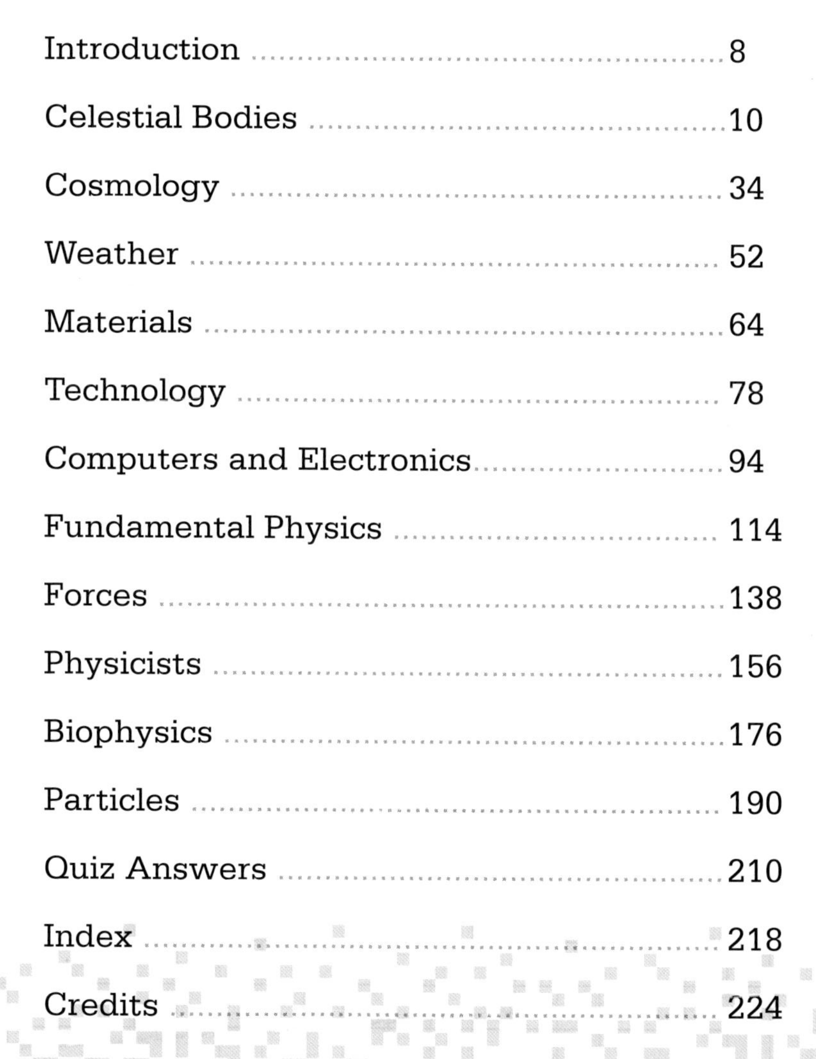

INTRODUCTION

Like it or not, the world around you is full of physics. Despite that fact, many of us go through our daily lives not realizing just how much of what happens is because of physics. For example, simply reading this book involves a great deal of physics: the atoms holding it and you together, the light letting you see the words, the electronics in your devices, the temperature in the air around you, gravity holding you and the book in place, and the movement of the earth below your feet.

In fact, many of the fundamental questions that people ask about the world can be solved by physics. In this book we provide the answers to more than a hundred such questions, which range from the esoteric such as "Does the philosopher's stone exist?" and "How heavy is a kilogram?" through to the down-to-earth such as "Can you really unlock a car using your head?" and "How does Wi-Fi work?" All of the answers are also crammed with interesting facts that make physics a lot more fun than you might remember from your schooldays—as well as helping you to be the smartest person in the room at the next dinner party.

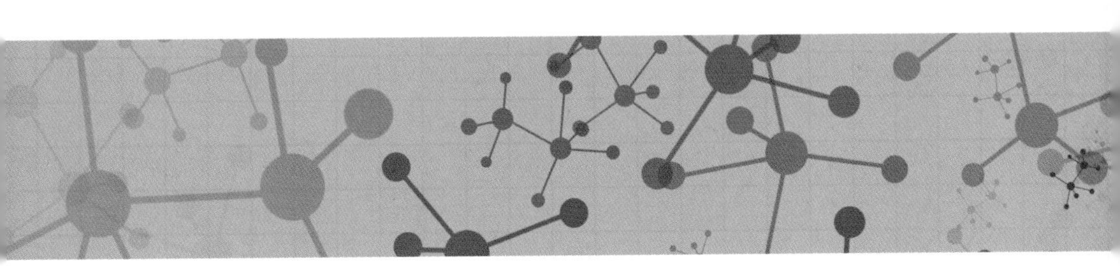

Throughout this book you will no doubt notice many links between the different questions, and that's one of the great joys of physics. The laws and rules around us are so interlinked that two wholly different questions—such as "Why do cars make that zooming noise?" and "What's at the center of the universe?"—might have more in common than you think.

Before we begin—a small warning. Physics is (and always will be) a work in progress. In the late 1800s, physics was considered a dying field of study with just a few loose ends to tie up, but then Albert Einstein turned everything on its head with his theory of relativity. So, in physics, answering a question often only raises more questions. In fact, the answer to each question here could fill a book on its own, but you will at least get a taste of the wonders of the complex world of physics.

WHERE IS THE SUN'S TWIN?
HOW DO SOLAR ECLIPSES OCCUR?
HOW HOT IS THE SUN?

WHAT MAKES STARS TWINKLE?

CELESTIAL BODIES

How do solar eclipses occur?

Solar eclipses are truly magnificent events—the sky falls dark during the day and a fantastic halo of light surrounds the Moon. They happen when the Sun, Moon, and Earth line up perfectly in an event called a "syzygy." The Moon moves into place between the Earth and Sun, blocking out the Sun's light and causing an eclipse.

Unique in the Universe?

If the Earth ever were to become part of some great intergalactic alien civilization, then it would probably be a tourist hot spot for its eclipses! The alignment of three objects in space is not rare, but a total solar eclipse certainly is. While there is some small variation due to orbital patterns, the Moon and the Sun look almost the same size in the sky when viewed from Earth. This is an incredible coincidence not seen anywhere else in our solar system, which leads to the striking ring of light seen during an eclipse. Astronomers believe it to be very rare and possibly even unique in the universe.

A Sign from the Heavens

Eclipses are difficult to miss, and they have been noted by humans ever since records began. In the past they were often seen as omens of great importance; many civilizations have tried to link the deaths and births of great historical and religious leaders or major events to eclipses. The ancient Greeks in particular put great stock in eclipses. The occurrence of an eclipse during the Battle of Halys in 585 BCE reportedly caused the troops to stop fighting, resulting in the swift arrangement of a peace agreement.

This one had been predicted beforehand by Thales of Miletus, suggesting that the ancient Greeks at least partially understood how and why eclipses occurred.

Other Types of Eclipses

If the Moon can get in the way of the Sun, then it makes sense that the Earth can get in its way, too. When this happens, the event is called a lunar eclipse. A lunar eclipse is often better known as a "blood moon," because as the Earth passes between the Moon and the Sun, only the red light from the Sun is able to reach the Moon's surface, making it briefly appear a deep, rich red color.

The orbit of the Moon around the Earth is not on a flat plane with the Earth's orbit around the Sun. This is why we don't get a solar eclipse every month. This wonkiness in the orbits means that we can sometimes get partial eclipses, where only part of the Moon passes between the Earth and the Sun. It's not only the Moon that does this. The planets Venus, Mars, and Mercury all at some point in their orbit pass between the Earth and the Sun. However, because the planets are much farther away from Earth than the Moon is, they aren't able to cause the same effect, as they appear very small. When a planet passes in front of the Sun, it is known as a transit—it is possible to see one using a specially equipped telescope.

Why are there no square planets?

Planets come in all sorts of sizes but only ever one shape: round. A planet can be made of rock, ice, or even gas, but the shape is always the same. This is because of the planet's own gravity pulling inward.

It's All in the Making

Gravity pulls everything toward everything else. The bigger the thing, the more pulling power it has. Gravity will always pull things toward the center of an object. This means that as a big object starts to form, it is going to have a very strong pull all over and will try to get as much stuff as close to the middle as possible—and the best shape for this is a ball. If a planet were a cube, then the corners would be farther away from the center than its faces, so the gravity would naturally pull on them until they flattened out, forming a sphere.

While this might make sense intuitively for gas planets, which are able to change shape easily, what about the more solid rock and ice planets? Their spherical shape is set when they are being formed. They start life as many small rocks or chunks of ice that are pulled together by gravity in the same way that clouds of gas are pulled together during the formation of a gas planet. However, as new solid material is pulled into the forming planet, this causes huge collisions. These crashes release a large amount of heat and the planets are turned into molten rock (in the case of rock planets) or liquid (in the case of ice planets) that gravity is then able to pull into the round shape we see today.

What's Not Round?

It's not just planets: Stars, black holes, and many other things in the universe are also made round by gravity. So what isn't? The short answer is anything that is smaller than about 370 miles wide, as it won't produce enough gravity to make itself spherical. This means that things like asteroids and comets can be all sorts of shapes. They are often roughly spherical, but they can come as long cylinders, strange lumpy blocks, or even (in the case of the comet 67P/Churyumov-Gerasimenko) sort of duck-shaped.

Is the Earth Really Round?

The Earth, like many planets, is not perfectly round. Its rotation and gravitational forces cause it to squish a little at the poles and bulge at the equator. The difference is about 17 miles between the Earth's width at the poles and the equator. The surface of the Earth is also not flat: The deepest point is the Mariana Trench at a little under 6.8 miles below sea level, and the highest point is the peak of Mount Everest at just under 5.6 miles above sea level.

The Earth is on average a whopping 7,918 miles wide (gravitational forces mean it varies by up to 13 miles). So what if we shrunk it down to the size of a billiard ball at just 2¼ inches wide? The difference between the width at the poles and the width at the equator becomes just 0.2 millimeters and the difference between the highest and lowest point on Earth a mere 0.1 millimeters. Official billiard balls have a roundness tolerance (as defined by the World Pool Billiard Association) of about 0.13 millimeters, meaning that the Earth is not quite as round as a billiard ball.

What makes stars twinkle?

You've heard the nursery rhyme and maybe you've even been out stargazing to see it, but the truth is that stars themselves don't actually twinkle. Stars look as if they are twinkling in the night sky because their light is being distorted as it passes through the Earth's atmosphere.

Light Distortion

The Earth's atmosphere is very thick and filled with many different gases. This causes light passing through it to be bent, split up, and bounced around, leading to a distorted image. The atmosphere above us is constantly shifting around, which changes how the light is being distorted all the time, making it get brighter and dimmer, which leads to the twinkling. The more atmosphere that the starlight has to travel through, the more it will be distorted. You may have noticed that stars on the horizon twinkle far more than those directly overhead. This is because if the light is coming from directly overhead, it comes straight down, whereas when you look toward the horizon the light from those stars is traveling at an angle, which means it will travel through the atmosphere for longer.

The Best Observing Conditions

Astronomers go to extreme lengths to minimize the effects of the atmosphere on their observations. Observatories are often built in deserts, where there is less water in the atmosphere above. They are also often constructed high up on mountains, where there is less atmosphere for the light to travel through. A more modern technique to limit atmospheric distortion involves firing a huge beam of light into the atmosphere. The light beam produces a stable and known pattern of light. What should be seen is matched with what is actually seen, and from this the current conditions can be determined. Images of stars and so on can then have the same corrections applied to them to get a more accurate image. Another way around the problem is to build satellite telescopes and launch them into space, where there is no atmosphere.

How hot is the Sun?

It might sound like a simple question and a quick search on the internet will give you an answer of around 10,000°F, but in truth it's not that easy. The Sun is a massively complex object with many different layers that each have a different temperature.

The Structure of the Sun

The Sun has a huge central core that acts as a furnace, fusing hydrogen atoms together into helium at a temperature of 28,300,000°F. Around this are several layers of hot, dense plasma that can be up to 12,600,000°F. The surface is a relatively cool 10,000°F; however, extending out beyond this, things can get even hotter! The solar winds and corona, which are caused by the Sun's massive magnetic fields, reach temperatures of up to 9,000,000°F. But even these temperatures pale in comparison with the solar flares, which are enormous jets of solar plasma that can reach 36,000,000°F!

Flying Close to the Sun

The Sun is very hot, so if you tried to fly by in a spaceship you'd get cooked, but what is the closest you can get? A space suit could keep you safe up to about 250°F and a spaceship up to nearly 5,000°F, so in a good spaceship you could get to about 1.3 million miles away from the Sun. The object that humans have put closest to the Sun is the Parker Solar Probe, launched in August 2018, which at its closest approach to the Sun will be 4.3 million miles away.

What are comets made of?

With their fantastic streaking tails, it's no wonder that comets capture the imagination of young and old alike. But what are these infrequent visitors to the night sky actually made of?

Big, Dirty Snowball

Comets are made of a mixture of rock, dust, ice, and frozen gases—a bit like a big, dirty snowball. The main body of a comet can be anything between a few hundred yards and several miles wide. The central part, called the nucleus, is made of a dusty rock surface covered in ice and frozen gases such as carbon dioxide, methane, and ammonia. Comets may also contain substances with more complex molecules, such as formaldehyde, ethanol, and potentially even hydrocarbons and amino acids. These are the chemicals that form the basis of life, and some theories suggest that perhaps it was comets crashing into the Earth during its formation that allowed life to start here.

Streaking Tails

The rather striking tail that can be seen trailing behind a comet is caused by the frozen gases on the surface of the comet melting in the heat of the Sun and being ejected off into space. Comets begin to form their tails when they get about 370,000 miles away from the Sun. Some of these tails can stretch out for millions of miles. It is worth noting that because a comet's tail is formed by the Sun and its solar winds, a comet's gas tail will always point away from the Sun, regardless of its direction of travel.

Landing on a Comet

On November 12, 2014, the European Space Agency's Philae module touched down onto the surface of a comet. While it wasn't a perfect landing, with the lander bouncing twice and then coming to a stop in a crack, it was the

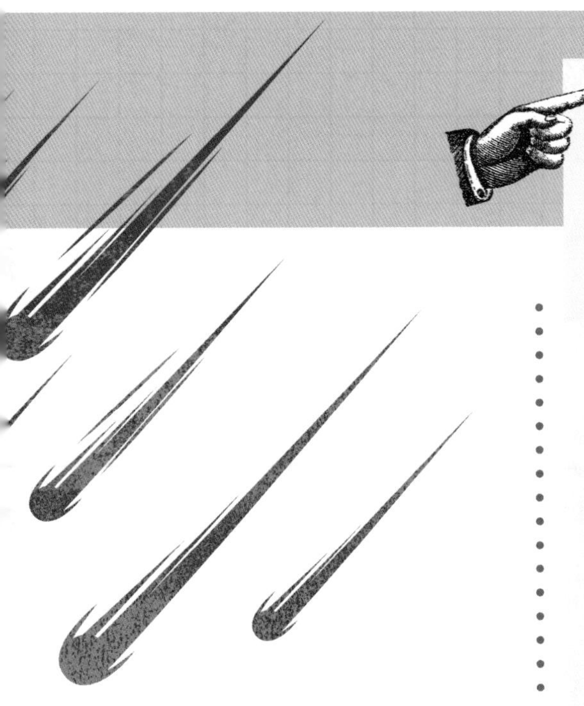

THE MOST FAMOUS COMET

Probably the most famous comet in the world is Halley's Comet. Its orbit around the Sun means that it comes into view of the Earth every 75 to 76 years. It has been recorded in history since antiquity, but it was only in 1705 that the English astronomer Edmond Halley realized that it was the same object that kept appearing in the sky and predicted when it would next appear (which was unfortunately 17 years after his death). The first record of Halley's Comet was in 240 BCE from ancient China, and it was also recorded by the Babylonians in 164 BCE and 87 BCE. Perhaps the most famous recording of the comet is in the Bayeux Tapestry, where it is displayed as an omen during the Norman conquest of England. The comet will next be visible in 2061.

first of its kind. Despite falling on its side, the lander was able to complete most of its scientific objectives, identifying a number of chemicals that hadn't been detected on comets before. It was also able to measure the temperature shift across the comet's 12.5-hour day from -292°F to -229°F and discover that the comet's surface is a 4- to 20-inch-thick crust of dust held together with ice, with a more "fluffy" porous rock layer beneath.

What is a shooting star?

A streak of light across the night sky—a shooting star! Maybe you feel like you should make a wish on it? While there's nothing stopping you, you should probably know that a shooting star isn't actually a star—it is caused by a piece of rock or space junk falling through the Earth's atmosphere.

A Shooting Star Is Born

When rocks or some of the millions of tiny pieces of space junk fall toward Earth, they get pulled in by gravity. They get faster and faster, reaching speeds upward of 23,000 feet per second. When they enter the Earth's atmosphere, they start to collide with air particles and other molecules, which produces friction, causing them to get very hot and burn up, forming large streaks across the sky. This usually happens at around 30–60 miles above the ground and the resulting shooting star lasts only a second or so. Shooting stars can happen all year round but are more common during meteor showers, when the Earth passes through a stream of debris left behind by something such as a comet.

Bigger and Bigger

Most shooting stars that you'll see are caused by something that's between the size of a grain of sand and a pebble, but they can get much bigger than that. Meteorites are the remaining bits of rock that can be found on the ground after a shooting star falls to Earth. To leave behind a rock that would fit comfortably into your hand, the original rock would have to have been about 3 feet or so in size. The largest meteorite ever found is known as "Hoba," after the farm where it was discovered. It weighs a whopping 66 tons and is a square roughly 10 feet wide and 3 feet thick.

Meteor Strike

As you can probably imagine, the impact of tons of rock traveling at thousands of feet per second can be incredibly dangerous. Fortunately, the most destructive events are very rare. The most well-known event of recent memory was the Chelyabinsk meteor in 2013, which was a 65-foot asteroid that was traveling at around 12 miles per second. As it passed through the atmosphere it turned into a huge fireball that could be seen over 62 miles away, and people close to it felt the intense heat as it exploded in midair. While spectacular, the event caused only minor damage to buildings.

Some meteors can be far more devastating. The Tunguska event, which took place on June 30, 1908, is thought to have been a meteoroid of between 200 and 620 feet exploding in the atmosphere. It blew up with a force 1,000 times greater than the atomic bomb dropped on Hiroshima and flattened an area of over 1,200 square miles, knocking down close to a hundred million trees. Fortunately, it occurred in a remote area of Siberia, so no people were hurt in the blast. One of the largest meteor impacts in the Earth's history caused the Chicxulub crater in Mexico. The meteoroid, estimated to be between 6 and 9 miles wide, landed 66 million years ago, and it is thought by some that the material thrown up into the atmosphere after the impact caused massive climate disruption, in turn triggering the extinction of the dinosaurs.

Is Pluto a planet?

Planets hold a special place in all of our hearts, but there are a lot of objects in our solar system and they can't all be called planets. So scientists have come up with a set of criteria for what makes a planet a planet, and despite the fact that many people have strong feelings about it, Pluto is not a planet.

How to Be a Planet

In August 2006 the International Astronomical Union set out the three conditions that an object must meet in order for it to be called a planet:

1. The object orbits the Sun.
2. The object is large enough such that its own gravity has made it spherical (see page 14).
3. The object has cleared the neighborhood around its own orbit of smaller objects.

Unfortunately for Pluto, it's the third point where it falls down. Pluto sits within the Kuiper Belt (which is like the asteroid belt), but sits beyond Neptune—this means that its orbit is full of other objects. Or if you look at it another way, Pluto is in essence just a particularly large asteroid in the Kuiper Belt. For this reason, Pluto lost its status as a planet in 2006.

OTHER EX-PLANETS

Between the 1600s and 1700s, many of the moons of Jupiter and Saturn were considered to be planets, until the Sun was accepted as the center of the solar system and they were reclassified as moons. In the early 1800s, the first asteroids—Ceres, Pallas, Juno, and Vesta—were discovered and as they orbited the Sun they were also called planets. Toward the middle of the 1800s, though, many more asteroids were discovered and they received their own category. Those asteroids that are spherical (like Ceres and Pluto) but have not cleared their orbits were reclassified as dwarf planets after 2006.

Why doesn't the North Star move?

The North Star has been used throughout the years as a clear point of reference for navigation, since it always stays in the same position in the sky, pointing north. The reason that it alone stands sentinel while all the other stars move? The axis of rotation (the imaginary pole around which the Earth rotates) is pointed directly at the North Star.

On the Move

The Earth's rotational axis doesn't stay perfectly still, it moves slowly over the years. The position that the axis points to moves in a circle over a period of about 26,000 years. While Polaris holds the title of the North Star at the moment, in antiquity a dark spot at the midpoint between the stars Alpha and Beta Ursae Minoras was used for the same purpose. Thousands of years in the future, stars like Vega, Alderamin, and Thuban will all have their turn as the North Star.

THE SOUTH STAR

On the other side of the planet, Sigma Octantis currently holds the title of South Star, but it is very dim and only barely visible on a clear night, so it is functionally useless for navigation. The constellation of the Southern Cross is used instead, as it points roughly to where the South Star would be. As with the North Star, the South Star changes over time as the Earth shifts. In some 60,000 years' time, Sirius, the brightest star in the night sky, will become the South Star.

Why do we only ever see one side of the Moon?

The Moon is an ever-present feature of our sky. Even looking at it with the naked eye, we can see that its surface is very distinctive. But if it's an orbiting body, then why does it always look the same? Why do we always see the same part of the Moon? It's because the Moon is tidally locked to the Earth.

Here's Looking at You

Tidal locking occurs when one object orbits another and, over a sufficiently long period of time, the gravitation tidal force (which is the pull of gravity between the Earth and the Moon) causes the rotation to slow down or speed up. What this means for the Moon is that gravitational pull from the Earth has made the rotation speed of the Moon slow down to a point where it orbits around the Earth every 27.3 days but also takes 27.3 days to spin on its axis. This means that the same side is always facing the Earth. It is not only the Moon that is affected by tidal locking; the locking process significantly helped to slow the rotation of the Earth to the 24 hours it is now from the 6 hours it once was. And the process is still ongoing—a year gets longer by 0.000015 second per year.

A Universe-Wide Phenomenon

Any two orbiting objects in space that are sufficiently large and close to one another will eventually become tidally locked. Usually, the smaller body becomes tidally locked to the larger one. Most of the major moons in the solar system are tidally locked to their planets, and the planet Mercury will likely become tidally locked to the Sun at some point in the future. The dwarf planet Pluto (see page 22) is tidally locked with its moon Charon, which is of a similar size. This means that if you were standing on Pluto's surface, not only would you always see the same side of Charon, but it would also always be in the same place.

The Twilight Zone

A planet that is tidally locked to a star would be a strange place indeed. It would have one side facing the star in constant day, which would likely become very hot. The other side would always be facing away and thus be in a permanent cold night. If the planet had enough material on it and some kind of atmosphere, the planet would likely be split into a frigid iced plane covering half of the planet and a baked desert on the other side. The only potentially habitable zone would be a thin band around the equator, where enough of the hot air would be able to keep the ice melted and something of a water cycle would be able to start. Life might just be able to make a home in such an eternal dim twilight.

THE DARK SIDE

Since the same side of the Moon always faces the Earth, it also stands to reason that the same side of the Moon always faces outward. This "dark side" of the Moon is, therefore, less shielded from meteor impacts, and because of this it is heavily cratered.

At what point does the sky become space?

Above your head right now is the sky, full of clouds, airplanes, and birds. Somewhere above that is space, with its stars, planets, and galaxies. What you might not have ever considered is exactly where one becomes the other. While there's no definitive answer, the point at which the sky becomes space is generally accepted to be the Kármán Line.

The Kármán Line

The Kármán Line is defined as sitting at 100 kilometers (62 miles) above sea level. It's based on a rough approximation of the point at which the air becomes too thin to sustain normal airplane flight through lift (see page 89). The idea is that anything that could fly beyond this point could also fly in deep space, making it a spacecraft. This definition is not exact. Due to the mixing of gases, air currents, and a whole host of other factors, the actual point where conventional flight becomes impossible would likely be different in different places and change over time—so 62 miles is just a rough estimate. It is also unlikely that any actual airplanes would be able to reach such a height, as they are not designed to operate on the very edge of the theoretical flight limit.

The Kármán Line is the value set by the French Fédération Aéronautique Internationale, which is the organization that sets the international standards for many things related to aeronautics. This is not to say it's used everywhere. NASA used to use the Kármán Line, but reduced its definition of the start of space to just 50 miles to be in line with the U.S. Air Force.

What's in the Atmosphere?

The sky isn't simply one great big single entity. It's a complex system generally defined by several layers that behave a little differently to each other:

The **troposphere** extends up to around 11 miles into the sky (but lower around the poles). This is the densest part of the atmosphere, and so it has the most going on. About 75 percent of all of the stuff in the atmosphere is concentrated here, and it's the region where most clouds form and airplanes fly.

Above this is the **stratosphere**, which reaches up to about 35 miles above sea level. This is where the ozone layer—which protects us from most of the Sun's harmful ultraviolet rays—sits, and it's also where the very highest clouds form. It has also been discovered that some birds are able to fly this high up.

Above this is the **mesosphere**. Here the air is very thin but just dense enough to be able to cause high-speed objects from space like meteors (see page 20) or space junk to burn up.

Finally, we reach the **thermosphere**, which extends to nearly 450 miles. This is where we find the Kármán Line, the auroras (see page 60) and an incredibly sparse collection of atoms that loosely make up the remains of the atmosphere. Beyond this there is only space.

Where is the Sun's twin?

Most Sun-like stars are born in pairs, and the Sun itself is likely no exception. However, when we look into the sky there's clearly only one sun up there. So either there never was a twin, or there was but now nobody knows where it is.

They Come as a Pair

A large number of all medium to large stars are born in pairs. This is because of the circumstances of their birth. Stars are generally made in enormous clouds of dust and gas that are pulled together by gravity. As the particles are pulled together, they rub against each other. As the stars get larger and larger, they manage to generate enough heat and pressure to ignite. Because this process takes place in truly enormous clouds, more than one star will form at the same time and the act of one star forming may even help other stars to grow.

Binary Star Systems

Over half of all the Sun-like stars that astronomers have observed come as a pair in what's called a binary star system. This is where the two stars orbit around some central point between them. These systems can also have planets around them. Systems with three or more stars in them can be very chaotic and prone to collapse or break apart. This is not to say it's not possible, but such systems often take the form of a binary star system with another star a long way out orbiting around both of them.

Where Can My Little Star Be?

The question remains: Where, then, is the Sun's twin? It's certainly not within the solar system, or even anywhere nearby. Searches conducted by powerful telescopes are hampered by gas clouds near to Earth that can block our view of regions of space—and despite our best efforts, nothing has been found. What has been found, though, is one of the Sun's other siblings. Identified through the composition of its elements, the star HD 162826 lies in the constellation of Hercules. The star isn't bright enough to be seen with the naked eye despite being slightly larger than the Sun, but research suggests that it was born in the same place that our own Sun was. So there may yet be hope of finding the wayward twin.

A STAR CALLED NEMESIS

The long-lost twin of our Sun is often dubbed "Nemesis," after the ancient Greek god of retribution. The name comes from the theory that the separation of the two stars and the twin star's later travels through the solar system caused it to throw huge amounts of comets and asteroids at the Sun. The original theories about the twin star suggested that Nemesis was a red or brown dwarf that still orbits the Sun at the very edge of the solar system. However, modern technology that is capable of surveying the farthest reaches of our solar system has failed to find anything. If there ever was a Nemesis, it has long since left the solar system.

What happens when all the planets align?

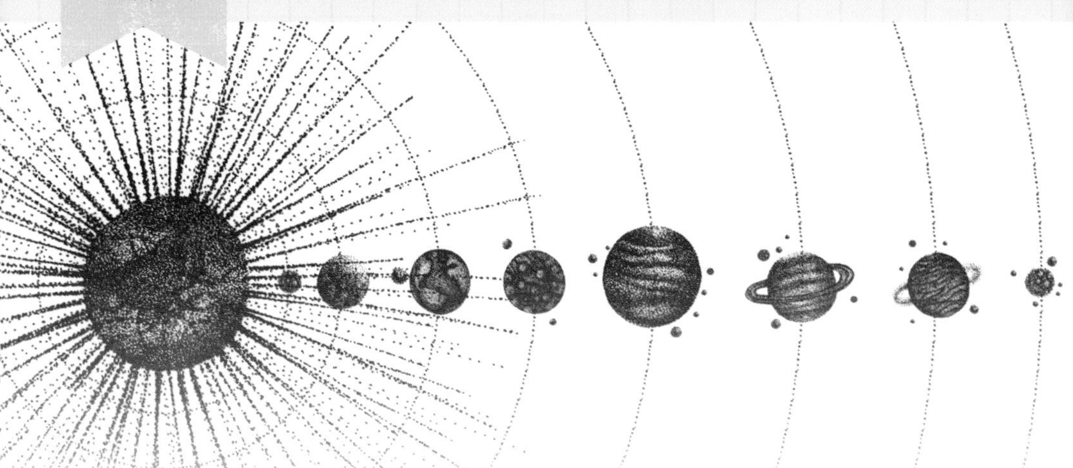

It's a staple of Hollywood films, prophecies, and astrology. When the planets in our solar system line up, there will be some sort of event of great power. Even the less fantastical accounts say that there will be strange gravitational effects that could cause chaos. But would it even be possible for all the planets to line up?

A Rare Occurrence

In fact, the planets in our solar system can never truly align. The planets orbit around the Sun at different angles, so it would never be possible for them to form a line. Because the planets orbit at different speeds, it is possible for all the planets to be in roughly the same patch of sky as seen from the Earth, but even at the point at which they are as physically close to each other as possible, they will still be fairly spread out. The next time that all of the planets will be visible in the sky at the same time won't occur until 2492, and even then they will be spread across a large area. Events like that only occur once every several thousand years.

What About Gravity?

Alarmists have suggested that planetary alignment could cause the effects of gravity to multiply, but this is simply untrue. The other planets in our solar system do have a gravitational effect on the Earth. However, this effect is incredibly tiny—so small, in fact, that even if it was somehow possible for them to line up, it wouldn't actually have any effect on our planet.

Where is the tallest known mountain in the solar system?

On Earth, we are mesmerized by the sheer enormity of our tallest mountains. The peak of Everest, for example, rises just under 5.6 miles above sea level. But in comparison with nearby planets and asteroids, our behemoths are nothing more than molehills. On Mars, there is a peak nearly three times taller.

Olympus Mons

As we don't have the information required to set a "sea level" equivalent on other planets, scientists calculate the size of non-Earth mountains from base to peak. Thought to be the tallest mountain in the solar system, Olympus Mons is situated in Mars's Tharsis Montes volcanic region. It measures a staggering 16 miles tall from its base. Olympus Mons's height was confirmed by NASA's *Mariner 9* space probe in 1971. We can look up and marvel at the scale of Everest, but Olympus Mons is so wide—it covers an area 388 miles across, which is roughly the size of Arizona—and slopes so gradually that if you were to stand below it, it would fill your view into the horizon.

Red Planet Plates

But why are mountains on Mars so much bigger? On Earth, tectonic plate movements prevent lava buildup beneath the crust, and this has caused the formation of smaller volcanoes in a chain—like the Hawaiian Islands. Mars's crust is made up of fewer plates and is inactive. Because of this, the position of the two main plates and lava hot spots has remained fairly consistent. This has allowed rising magma to erupt on the same part of the planet multiple times, where it cools and hardens, slowly creating a vast volcano.

How many moons does Earth have?

Moons are celestial bodies orbiting around a larger body, such as a planet, which is presumably orbiting around a star. Other planets, such as Jupiter, have numerous moons. At last count, there were 67 of them orbiting the gas giant. That makes our Moon seem rather lonely by comparison, but in fact it isn't alone up there.

Temporarily Captured Objects

The Moon is 2,175 miles across and has been orbiting Earth for over four billion years. However, there are thousands of other temporarily captured objects, or "mini moons,"caught in Earth's orbit, some of which measure a few feet across, and most of which are even smaller. In 2006 a sky survey conducted by the University of Arizona found a mini moon the size of a car. It was named 2006 RH120 and left Earth's orbit to resume orbiting the Sun less than a year later. In March 2011, a team of scientists used a supercomputer to calculate that at any given time there is at least one asteroid orbiting the Earth that measures 3 feet across. These asteroids don't orbit the Earth in neat circles—the twisty path they follow is a result of their being pulled between the gravity of Earth, our Moon, and the Sun.

More Moons?

It is possible that in the past Earth had a second large moon. That might explain the strange terrain on the far side of the Moon, which could be the result of our Moon crashing into another. Moons come and go—Mars currently has two large moons, but one of them is headed straight for the planet, where it's expected to crash in the next 10 million years. There's also a chance our planet could acquire a second large moon in the future.

CELESTIAL BODIES

Test how much you've learned about the Earth and the solar system with this quick quiz.

Questions

1. What does the atmosphere do to light coming from distant stars?
2. What needs to move in front of the Sun for an eclipse to occur?
3. What size do objects in space need to be in order to become round like the Earth?
4. Just how hot is the surface of the Sun?
5. What is the name of the comet featured on the Bayeux Tapestry?
6. They're not falling stars, so just what are shooting stars really?
7. What is the name of the belt of asteroids, comets, and other stuff that Pluto sits in?
8. What is the southern equivalent of the North Star?
9. The length of a day on Earth has slowly changed over time. How long was it just after the planet had formed?
10. When will the planets in our solar system next roughly align?

Turn to page 212 for the answers.

WHY ARE GALAXIES FLAT?
WHAT'S THE HOTTEST PLACE IN THE UNIVERSE?
WHAT'S THE OLDEST THING WE'LL EVER SEE?
HOW BIG IS THE UNIVERSE?

WHAT MAKES DARK MATTER DARK?

COSMOLOGY

WHAT'S THE BRIGHTEST THING IN THE SKY?

Why are galaxies flat?

Galaxies are enormous clusters of billions of stars. They are truly awe-inspiring to look at (though it can require powerful telescopes). There is a huge amount of variety out there, but they all have something in common. All galaxies are flat, and they are all flat because they all spin.

What Exactly Is a Galaxy?

Galaxies are systems of billions of stars, trillions of planets, thousands of black holes, neutron stars, pulsars, and much more besides. They're all held together by gravity. There could be anywhere between 200 billion and 2 trillion galaxies in the universe, and while they vary in size, they can be up to 300,000 light-years (1.7 quintillion miles) across. Pretty much everything in the universe exists inside galaxies. We might think of the space between the planets or even stars as empty, but compared to outside a galaxy, it's practically stuffed to the brim with matter.

Galaxy Formation

Galaxies form in the same way as everything else in space—gravity pulls things together. The first galaxies would have been truly titanic gas clouds many times the size of the galaxies we see today. These condensed, forming stars, planets, and everything else. As they did so, all of these newly created stars were also pulled together by gravity. This process pulls a lot of stuff into the middle of the growing formation, which creates a galactic center where most of the stars are concentrated, making it appear as a big bright bulge. Much of the rest of the material then becomes spread out around the center in a wide rotating disk. There is also a "halo" of disparate stars and other stuff that surrounds the galaxy in a light, spherical cloud of matter.

Whirling Silently in Space

As a galaxy is forming and pulling in matter, it starts to spin. In the same way as when a star or a planet forms (see page 14), the act of pulling something toward it starts a rotation, which then makes other things nearby begin to rotate in the same way, until you have a giant ball of matter that is spinning as it condenses down. Because of this spin, the centripetal forces push outward along the area perpendicular to the motion of spin. This means that while most of the stars and other matter are pulled inward along a particular axis, they're pushed out and flattened—like a chef spinning pizza dough in the air. Galaxies are almost like ringed planets, only where the rings are bigger than the central part, and rather than just a bit of rock and ice they are instead made of billions and billions of stars.

What are black holes?

You've heard about black holes; they're dangerous, exciting, and open the door to a million adventures or future technologies. But what are they really? This is not an easy question to answer, and they're still far from fully understood, but black holes are basically massive objects that have such great gravitational pull that nothing can ever escape.

How to Make a Black Hole

If you take some matter and let gravity pull it together, you might get something like Jupiter—just a big ball of gas. If you have even more matter, you get a star. If you let that star undergo various processes, it may end up as a very dense object like a neutron star (a thimbleful of neutron star weighs more than Mount Everest) or a white dwarf, where it's full up to the point where the strongest atomic forces pushing out are only just in balance with the gravity pulling in. To make a black hole, you just add some more mass.

When you reach a certain amount of mass in a small enough space, then physics starts to get really weird. With enough mass crushing in under gravity, it is eventually enough to break down and cause all of the mass to become a single point. Just an infinitesimally small dot that contains an enormous gravitational pull. This is a black hole.

The Event Horizon

Gravity pulls us toward massive objects, and black holes are about as massive as things get. They have so much gravity that it can become impossible to escape. While the black hole itself is only a small dot, it has around it an "event horizon"; this is the point at which its gravity becomes so strong that even light can't get away. The event horizon means that when we look at a black hole we see a dark, empty circle in space, as it eats anything that might shine there—hence the name of "black hole."

Beyond the Horizon

Let's say you were in a spaceship that could travel faster than the speed of light, so you decide to nip into an event horizon and see what is going on. This is what you would find. As you cross over the event horizon, nothing. There is no change that you can see or even initially detect. Bored and likely disappointed, you turn your ship around and power up the faster-than-light engines and blast away. Or rather, you blast deeper into the black hole! The black hole's gravity is so strong that it actually causes space to fold in on itself, making all directions of space point toward its center. You are now trapped, and as you get closer to the black hole itself the gravity at your legs is stronger than at your head, as your legs are closer to the center, so you start to get stretched out in a process called spaghettification. This will all take some time. Albert Einstein showed that space and time are really the same thing, and because space is messed up, so too is the time. In fact, as you get closer to the center of the black hole, time will start to slow down more and more. As you look out into the universe you will see stars being born and dying before your very eyes as the star's time moves faster compared to you, until eventually at the black hole itself time may just stop. (We're not sure on this; physics gets really messy in a black hole, so it's a best-guess scenario.)

What's the hottest place in the universe?

There is a theoretical maximum temperature known as the Planck temperature, which in Fahrenheit is at 2.5 with 32 zeros after it; however, nothing has ever come close to this, and it is unlikely anything ever will. The hottest place in the universe is inside the Large Hadron Collider (LHC) at the CERN facility on the border of France and Switzerland.

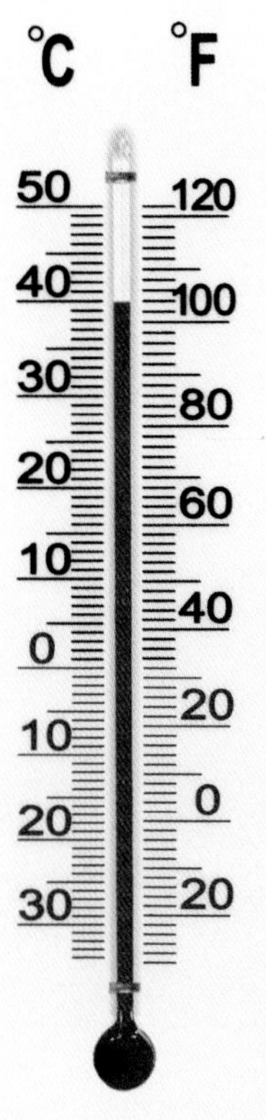

The Hottest Thing

The LHC works by accelerating particles up to near the speed of light and crashing them into each other. By smashing two gold particles into each other, it is possible, if only for a fraction of a second, to reach temperatures of 7,200,000,000,000°F. This is twice the Hagedorn temperature, which is the point where normal matter breaks down. Scientists exploit this fact and use it to split atoms apart into various components, which they can then examine within the LHC.

The Hottest Natural Thing

There are a lot of hot things out there in space. Our own Sun can reach several millions of degrees (see page 17) and even larger stars can get nearly ten times as hot. There is also a cluster of galaxies called RXJ1347 which are colliding and have heated up to over 540,000,000°F.

The hottest thing, however, is in the core of a supernova (see page 48). Supernovas are enormous explosions caused by collapsing stars. The enormous pressure within a supernova is able to produce temperatures in excess of hundreds of billions of degrees.

What's the coldest place in the universe?

The coldest anything can get is 0 degrees kelvin (-459.67°F). This is the point where everything stops moving, even atoms. Quantum physics tells us that it is not possible to reach this temperature, so what is the closest thing? The coldest temperature ever recorded was on a tiny piece of metal in Colorado.

The Coldest Thing

In 2016 researchers at the National Institute of Standards and Technology in Boulder, Colorado, took a tiny piece of aluminum (only 0.02 millimeters wide) and, using a special laser technique called "sideband cooling," managed to cool it down to just 0.00036 degrees kelvin.

Low temperatures make it a lot easier to study much of physics. Temperature is a measurement of the movement of atoms within an object. At lower temperatures everything is moving around a lot less and unexpected things are less likely to happen, so there are a lot of reasons for continuing this kind of work. Research teams in Gran Sasso, Italy, and Lancaster, England, are among those who have previously held the record for coldest place in the universe and scientists may, in the future, be able to create even colder temperatures still.

The Coldest Natural Thing

The coldest naturally occurring thing in space is the Boomerang Nebula. It's a star at the end of its life that is throwing out a huge amount of frigidly cold gas. This gas is being forced out at such a rate that it expands rapidly, causing it to become supercooled. This lowers the temperature of the space around it down to just 1 degree kelvin—colder even than empty space, which sits at 2.7 degrees kelvin.

What makes dark matter dark?

You may have heard about dark matter. It's one of the greatest mysteries in our universe, and even its name alludes to the things still to be learned about it. Dark matter is so called because it doesn't give off any light, and that's about all we know about it.

Something Out There

There is a lot of dark matter out there. In fact, there is more than five times as much dark matter as there is normal matter. But if we've never seen it, and we don't know what it is, how do we know that there is any dark matter at all? Let alone how much of it there is.

In 1933 a scientist named Fritz Zwicky was working on calculations of a galactic cluster millions of light-years away. He determined the mass based on how bright it was and then did his calculations. He quickly realized that they were moving faster than expected. The only explanation was that there was missing matter not contained within the brightness, some 40 times more than what could be seen. He therefore dubbed this "dark matter." It took a while, but it became clear that it wasn't just this distant galactic cluster that was missing matter. Every cluster was missing mass, and it turned out that even our own galaxy was, too. The solution to all of these problems was dark mass—some invisible form of mass that sits on the edge of galaxies and galaxy clusters, forming a huge web throughout the universe.

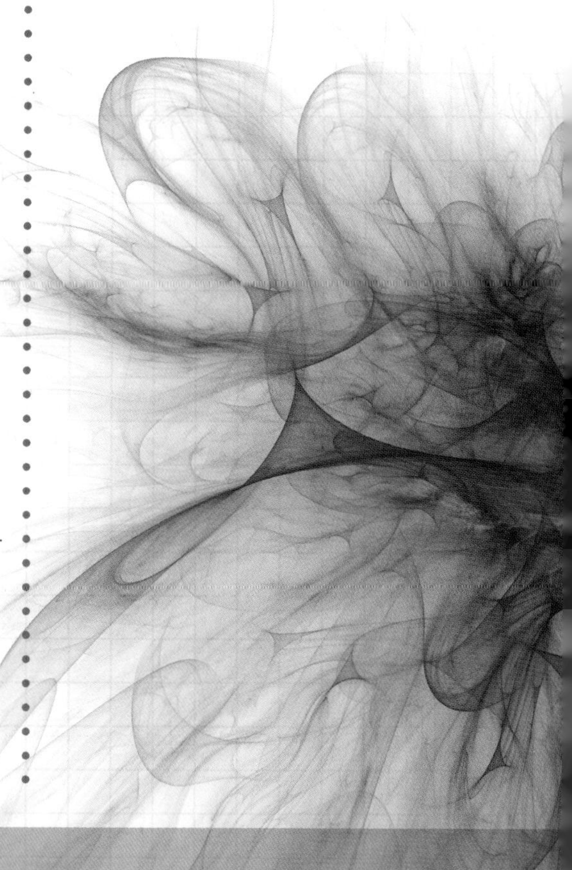

Mysterious Matter

We don't know what dark matter is, but there are many theories as to what it could be. An initial suggestion was that there are a lot more black holes, planets, and brown dwarf stars out there than we thought. These are essentially invisible to modern technology and provide some mass. However, even the wildest estimates don't get close to covering the missing mass. The leading theory for what dark matter might be is weakly interacting massive particles (WIMPs). These would be tiny particles that interact with the universe only through gravity and the weak nuclear force (the weakest of the four fundamental forces), making them even harder to detect than neutrinos (see page 208). Some of the more imaginative theories suggest that perhaps gravity doesn't work in the way we think it does on such large scales, or that perhaps gravity is somehow coming from other dimensions. While these ideas should not be discounted, there is little compelling evidence to believe that they are possible.

DARK ENERGY

Dark matter isn't all that's missing from our understanding of how the universe works. Calculations on the expansion of the universe throw up the question of unknown energy that seems to fill empty space. This has been called dark energy for its mysterious origin and makes up nearly 70 percent of everything in the universe. Scientists know even less about this than they do about dark matter.

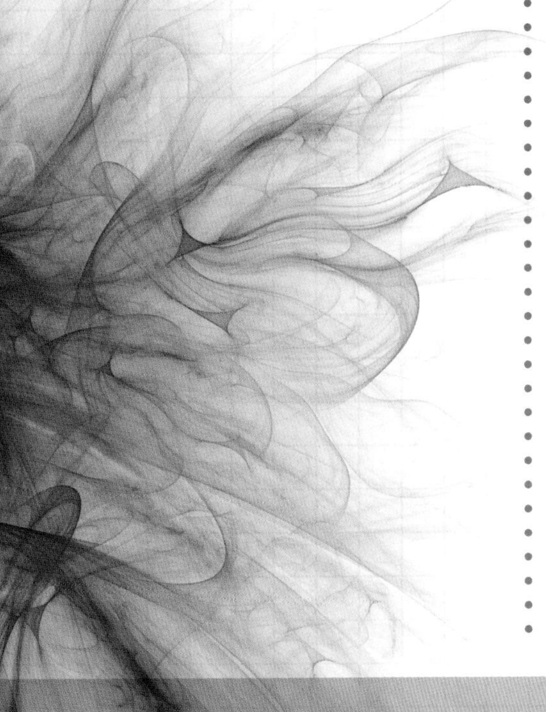

How big is the universe?

As Douglas Adams once famously wrote, space is big. Really big. The observable universe is a whopping 46 billion light-years across, but the universe itself could be much bigger.

Universal Expansion

Ever since the big bang, the universe has been getting bigger. The big bang may even be more accurately called the big expansion. The universe went from an infinitesimally small dot and got bigger very quickly. In the inflationary epoch, it went from about 1 nanometer across to nearly 11 light-years in less than a hundred billion billion trillionths of a second (1 with 32 zeros in front of it). After this rapid expansion, it slowed down dramatically but continued getting bigger at something like the speed of light. The universe is about 13.8 billion years old, so it would make sense that the edge of the universe is 13.8 billion light-years (plus that initial expansion of 11 light-years) from the middle, and then if that goes both ways the total width of the universe is about 27.6 billion light-years across. But it's not quite that simple.

It Keeps Getting Faster

In 1998 two teams studying distant supernovas realized that not only is the universe expanding but the rate of that expansion is increasing. It also became obvious that this was a relative effect: Everything is moving away from everything else, and the farther apart objects are from one another, the faster they move away. This places a limit on how far we can see. Objects far enough away from us could move faster than the speed of light farther away from us, so we can never see them. Calculations cap the farthest we can see at a universal width of 46 billion light-years. The universe, though, likely continues beyond this and could, in theory, be infinitely large.

What is at the center of the universe?

If the universe expanded from a single point, there's a question that remains. What's there now? A black hole? An ancient race of the first aliens? Rather disappointingly, the answer is that there is nothing at the center of the universe, because there isn't really a center of the universe.

Universal Expansion (Again)

The universe expanded from a single point, yes, but not in the same way, say, a bomb might explode. It wasn't an expansion in space, but rather an expansion of the space itself. The universe is (probably) four-dimensional and this means that when it expanded it did so in the fourth dimension, and the expansion we see in our three dimensions is something of a by-product of this. This means that everything is expanding away from everything else; there is no central point. And the point where this expansion started only exists as a fourth-dimensional point, which doesn't exist here in our universe for us to ever visit or examine.

The Problem of 3-D Thinking

Humans are three-dimensional beings living in a three-dimensional world (or four-dimensional, if you include time; see page 127), so trying to up the number of dimensions we think about often just doesn't work. Instead, it's better to think down in terms of dimensions using the well-loved example of a balloon. Think of a two-dimensional universe drawn on the surface of a balloon. Dots and swirls indicate galaxies and other matter. When you blow air into a balloon, it expands in three dimensions but also in two. The galaxies get farther away from each other and also expand outward, like in the universe. The center, however, is left inaccessible from the two-dimensional universe much like here in our own, where the center could only be found in the fourth dimension.

How will it all end?

Nothing lasts forever. Probably even the universe will one day come to an end. You can take solace in the fact, though, that scientists have estimated that the universe's eventual death won't happen for billions if not trillions of years. Nobody really knows how or when the universe will end, but we have lots of different ideas.

Heat Death

Ever since the big bang, the universe has been winding down. Entropy (see page 132) means that the energy in the universe is constantly being spread out. The heat death theory holds that eventually this will reach a point where everything is so spread out that everything is doing nothing. In the heat death, all of the elements made by the big bang will become inanimate. The stars will go out, with nothing to replace them. All of the galaxies will cool down into inert balls of matter, and even black holes will evaporate into nothing. The whole universe will become a dead sea of stuff where nothing ever happens or changes.

FALSE VACUUM THEORY

There are also other, larger, more theoretically based ideas about the end, such as the false vacuum theory. It states that the universe itself may be unstable and that it is only one possible configuration of a larger system that could suddenly alter. If it did suddenly alter, it would cause everything in the universe to be destroyed instantly.

Big Crunch

We don't know why or how the universe is expanding, so it follows that it might one day stop getting bigger and go into reverse. The big crunch hypothesizes that the universe may begin to collapse in on itself; the galaxies will be pulled into each other rather than being thrown apart. As the universe gets smaller and smaller, not only will galaxies begin to collide but so will everything else, as it is crushed into a smaller and smaller space, essentially re-creating the big bang in reverse, until there is once again nothing. Some scientists suggest that this may then cause a new big bang, starting the process all over again, in the "big bounce" theory.

Big Rip

We know that the universe is expanding and that the rate of expansion is getting faster, so perhaps this will just keep happening. This could eventually reach the point where the universe is expanding so quickly that even over small amounts of space it is expanding faster than the speed of light. This would result in the amount of the observable universe shrinking ever smaller as the universe beyond it speeds away from us. Eventually, this could result in even individual atoms or components of atoms being separated from everything else, until just the single smallest possible parts of the universe are sitting alone in their own practically infinite space. A lot of the uncertainty over why or how the universe will end comes from the unknown influences of dark matter and dark energy, so until we know more about them, we can never be sure.

What's the brightest thing in the sky?

There are a lot of shiny things in the night sky: stars, planets, the Moon. There are even more exciting things like accretion disks around black holes (stars being pulled apart and stretched out), nebulae, and more. But the brightest thing in the night sky is a supernova.

How to Make a Supernova

There are two main types of supernovas, type Ia and II. They both involve a stellar core collapsing and then undergoing thermonuclear detonation, but exactly how this happens varies a little. A type Ia supernova comes from a binary star system, usually a white dwarf and a red giant. The white dwarf is able to leech material off the red giant in a vampiric interaction that makes the white dwarf heavier and heavier. When its mass reaches the Chandrasekhar limit of just over one and a half suns, its core collapses and then goes supernova. When a very large star collapses, the remaining core could already be large enough to break the Chandrasekhar limit. If this is the case, then it's a type II. Because it doesn't always happen at exactly the same point, there is a lot more variability in how big and luminous the explosion is, whereas type Ia supernovas are all very similar.

Going Supernova

Supernovas are exceedingly bright. Supernovas explode with the energy of ten billion billion trillion atomic bombs per second. A single star going supernova can be as bright as the rest of the galaxy it's in for its short duration. This means that it is possible to see individual supernovas occurring in far-flung galaxies millions of light-years away. They are so bright that it can be hard to get a real grasp on it. If you had the choice between our own Sun going supernova or the most powerful bomb ever created going off in front of you, then you'd want to choose the bomb every time. (Needless to say, neither of these options would end well for you.)

BETELGEUSE, BETELGEUSE...

Betelgeuse is the ninth-brightest star in the night sky. It sits at the right armpit of the constellation of Orion and is a relatively easy star to spot. It is a truly massive red giant star a billion times larger than the Sun and up to 150,000 times brighter. Betelgeuse is primed to go supernova (it possibly already did, but if that's the case then the light from it is yet to reach us) very soon. The problem is that in astronomy terms that means it could explode tomorrow, or in a million years' time. If it does explode in our lifetime, then for a couple of weeks it will be almost as bright as the full moon, making it visible during the day, before winking out forever.

Visiting Stars

Supernovas have been recorded throughout history. Ancient Chinese astronomers made recordings throughout the centuries, and ancient markings by Native American tribes in Arizona suggest that they saw them even before the Chinese did. Many civilizations have observed these temporary stars in the night sky and ascribed various religious meanings to them. But in the late 1500s, the appearance of a new star prompted astronomers like Tycho Brahe (see page 170) to question whether the skies were really as immutable and unchanging as had been previously thought.

What's the oldest thing we'll ever see?

Light takes time to travel, so the farther away we look, the farther we're looking into the past. The night skies above us are filled with the light of stars that are long dead. So what's the farthest we can look back with this?

The CMB

The oldest thing we can see is the cosmic microwave background (CMB).The CMB was formed by the recombination event about 380,000 years after the universe began. At this time, the universe cooled enough to allow electrons and protons to form together to become the first atoms. These new atoms weren't able to absorb all of the energy that was around, so for the first time the universe became visible, as electromagnetic waves such as light were able to travel. The CMB was originally released at a temperature of about 5,500°F as light waves; however, as it has cooled over billions of years, it is now just -455°F and has been stretched out into microwaves. The CMB is in every direction throughout space and is what gives space its temperature.

A Strange Hum

There were some theories predicting the CMB in the mid-twentieth century, but it wasn't until 1964 that it was found. American radio astronomers were working on a new instrument that was designed to look deep into space and quickly discovered a strange hum. The hum came from whatever direction they pointed the telescope, so they assumed there was something wrong with the equipment. After checking all the cables, removing possible sources of electrical interference, and even removing a family of pigeons that had moved into the antenna and cleaning up their mess, it was still there. The hum was eventually linked to the relevant modern theories, and the astronomers realized that they had discovered an echo of the big bang.

COSMOLOGY

Are you a master of the universe? Think you know everything there is about space? Prove it by answering this quiz.

Questions

1. What tasty food gets its distinctive shape in the same way as galaxies?
2. What is the boundary of a black hole called?
3. At a scorching 540,000,000°F, what is the hottest naturally occurring object in the universe?
4. What is the temperature of empty space in kelvin?
5. What percentage of the universe is dark energy?
6. How old is the universe?
7. How many spatial dimensions does the universe probably have?
8. What is the name of the idea that the universe may collapse in on itself?
9. What does CMB stand for?
10. What limit do star cores need to overcome to go supernova?

Turn to page 212 for the answers.

CAN IT BE TOO COLD TO SNOW?

WHY DO YOU SEE LIGHTNING BEFORE YOU HEAR THUNDER?

WHY DOES IT ALWAYS RAIN AFTER A ROCKET LAUNCH?

COULD A BUTTERFLY REALLY CAUSE A TORNADO?

WEATHER

WHY DO HURRICANES SPIN?

Can it be too cold to snow?

You've probably heard at some point someone saying that it's too cold to snow. This doesn't make sense intuitively, as it's the cold that makes it snow in the first place. This is something of a half-truth. It can't be too cold to snow, but as it gets colder, snow is less likely.

The Snow Sweet Spot

Most snow occurs when the temperature is between 12°F and 32°F—note that this is the temperature at cloud level, not necessarily on the ground, where it may be slightly warmer or colder. Water vapor in the air begins to freeze at these temperatures, forming snow. As the atmosphere gets colder, the air is able to hold less and less water vapor. At temperatures below -4°F there is very little water around to crystallize into snow, and snow formation becomes pretty much impossible. So while it's not impossible for it to be too cold to snow, that's not because of the temperature—it's because it's too dry, though this in itself is caused by the low temperature.

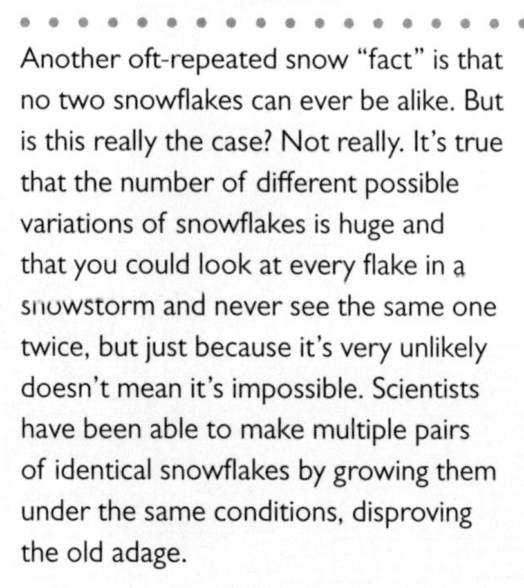

NO TWO ALIKE

Another oft-repeated snow "fact" is that no two snowflakes can ever be alike. But is this really the case? Not really. It's true that the number of different possible variations of snowflakes is huge and that you could look at every flake in a snowstorm and never see the same one twice, but just because it's very unlikely doesn't mean it's impossible. Scientists have been able to make multiple pairs of identical snowflakes by growing them under the same conditions, disproving the old adage.

Why can't you reach the end of a rainbow?

Whether they were in search of a pot of gold or simply wanted to see what it looks like, many people have chased after the end of a rainbow. Sadly, no one has ever succeeded. It is impossible to reach the end of a rainbow because a rainbow is not a real thing, it's just an optical illusion.

How Rainbows Form

White light (like the kind that comes from the Sun) is made up of a mixture of all the other colors of light. When white light passes through a shaped transparent object such as a glass prism (a pyramid-shaped piece of clear glass), the light splits up into all of the different colors.

Rainbows form when the Sun is behind you and some kind of water droplets (from rain, for example, or a spraying hose) are in front. The light enters into individual water droplets and splits into all the different colors of light. However, from each water droplet only one color of light reaches your eye. Light is split up and sent in lots of directions, so which color light comes out of each drop depends on where it is. When there are millions of water droplets in the air all at the same time, the different colors of light from all the droplets builds up together like a dotted painting to create the image of a rainbow.

Somewhere Over the Rainbow

Because it's an effect of the light being reflected at you, when you move, the raindrops—which remain in the same position—reflect different colors of light, meaning that the rainbow also seems to move. It's not a physical thing and everybody sees a slightly different rainbow. So try as you might, you'll never manage to reach the rainbow.

Why do hurricanes spin?

Hurricanes are hugely destructive storms that can cause devastation across a wide area. If you've ever seen a satellite image of one, you've probably noticed that they spin. The spin of hurricanes (also known as cyclones when they form in the South Pacific and Indian Oceans, or typhoons in the Northwest Pacific) is due to the Coriolis effect.

The Coriolis Effect

To understand the Coriolis effect, we must first think about a basketball. Imagine placing two stickers on it, a blue one on the middle of the ball and a red one very close to (but not on) the top, then placing your finger on the bottom and giving it a spin like a pro. Let's say the ball rotates on your finger once per second. If you think of the red sticker on the top of the ball, it will be moving fairly slowly; however, if you look at the blue sticker in the middle of the ball, it will be moving much faster, as it has a longer distance to travel. This is the basis of the Coriolis effect—that a solid rotating object (like a basketball or the Earth) will have points rotating at different speeds.

As the Earth spins, this naturally moves the air in one direction. Much like the basketball, the Earth moves faster at

the equator than at the poles. This means the air above the equator moves faster, too. Because of this difference in speeds, if something was moving through the air from the North Pole down in a straight line toward the equator, the faster-moving air toward the equator would cause it to veer off to its right (west). If something were moving up from the South Pole, it would curve to its left (also west). On the Earth, therefore, this creates a clockwise motion in the Northern Hemisphere and a counterclockwise motion in the Southern Hemisphere.

Spinning Hurricanes

Hurricanes are huge areas of low pressure in the atmosphere within which systems of clouds and air currents form. At the equator the Sun heats up the sea, leading to the evaporation of a lot of water, which forms the initial clouds; as evaporation from the sea continues to drive warm, moist air upward, these clouds build and develop to the point of near-constant thunderstorms as they form and move. Hurricanes are large enough and initially slow enough to be affected by the Coriolis effect. It causes them to spin as they move away from the equator, allowing them to reach very high wind speeds of above 75 miles per hour. The direction that hurricanes spin is determined by which hemisphere they're in. If hurricanes reach land, they can cause massive devastation thanks to their high-speed winds, massive amounts of rain, and "storm surge," an effect caused by the hurricane pushing seawater onto land.

TOILET FLUSH

There's an old urban legend that says that because of the Coriolis effect, when you flush the toilet, the water in the toilet bowl spins either clockwise or counterclockwise, depending on which hemisphere you are flushing in. This is not true. On such a small scale, it is the shape of the bowl that causes the direction of spin, not the Earth's rotation.

How do firenados form?

Firenados (more accurately called fire whirls) are towering columns of spinning fire that are as dangerous as they are mesmerizing. They can be caused by any large-scale fire, usually a forest fire, whether naturally occurring or otherwise. They form when the winds combine with rising heat from fires to create a towering inferno.

Whirlwind to Fire Whirl

Whirlwinds are simply spinning columns of air. Interruptions to normal wind patterns, which can happen because of sudden temperature changes or even changes in the height of the land below, can cause a column of spinning air to form, which is driven along by the normal wind. These can form anytime, anywhere. Fire whirls start in the same way as whirlwinds but they are fueled by fire, which adds extra instability at ground level, helping them to form. The column of air draws the fire up inside it. Fire whirls are usually only 30–160 feet tall but can be in excess of half a mile!

DEADLY FLAMES

Fire whirls can reach temperatures above 1,800°F and internal wind speeds greater than 90 miles an hour. As they move across the land, they spread fire and the resulting burning debris can rain down for miles around. A large fire whirl occurred after an earthquake in Kanto, Japan, in 1923, which caused a large number of fires. In its 15-minute life span, the fire whirl caused the deaths of thousands of people.

Why do you see lightning before you hear thunder?

Thunderstorms can be exciting and scary: a brilliant flash of light followed by deafening thunder. It's always in that order (unless it happens directly above you). But why? It all comes down to the speeds at which sound and light travel through the air.

First the Flash . . .

Lightning strikes are essentially no different than the sparks of static energy that might give you a shock when you touch something metallic (see page 70). Inside storm clouds, there are many particles of ice that bump into each other, causing static. This builds up a huge amount of electrical charge. The top of the cloud becomes positively charged and the bottom becomes negatively charged. This electricity mostly discharges from cloud to cloud as sheet lightning, but if enough charge builds up, it can overcome the natural boundary of the air and discharge as a lightning bolt. Lightning bolts are over 49,000°F and contain a billion joules of energy, enough to power the average home for more than ten days. Light travels at about 983,267,717 feet per second, so the lightning flash is picked up by your eyes almost instantaneously.

. . . Then the Sound

Thunder is the sound made by lightning. As it strikes, it causes the air to heat rapidly and thus the air around it to expand, which causes a loud shock wave similar to sonic booms from high-speed aircraft. Sound travels at about 1,115 feet per second, which is significantly slower than the speed at which light travels.

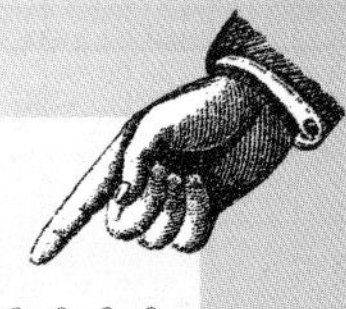

COUNTING SECONDS

From this, it's possible to figure out how far away a thunderstorm is. Every five seconds between seeing lightning strike and hearing thunder counts for a mile. So if you count ten seconds, the storm is two miles away.

What causes the northern lights?

The world is full of dazzling sights, but perhaps one of the most captivating is the northern lights, or the aurora borealis—waving curtains of multicolored ribbons that dance and light up the sky. The northern lights are caused by solar radiation impacting the Earth's atmosphere.

Aurora Borealis

Aurora is the Roman goddess of the dawn, while *borealis* is Latin for "north," so the literal meaning of this name is the northern dawn lights, though they are not limited to appearing at dawn. The Sun ejects a large number of particles out into space, and some of them reach the Earth. When these particles hit the Earth's atmosphere, they can bump into the atoms of various gases that make up the atmosphere. This bumping causes the atoms to release colored light; when enough of this happens at the same time, it forms an aurora. Green is the most dominant color in the aurora as it is released by nitrogen, which is the most common gas in the atmosphere. Oxygen releases red and sometimes blue light. Various forms of invisible ultraviolet and infrared light are also released by the northern lights.

Why Northern?

The northern lights are only seen in the more northerly parts of the world (hence the name). This is because the solar winds are magnetically charged, so they are drawn up to the north pole by the Earth's magnetic field and then pulled down into the atmosphere there. That said, the more active the solar wind, the farther south the lights can be seen. There are also southern lights (the aurora australis), which are made in the exact same way. These are less commonly seen because they are very far south—even in the most extreme solar winds, they can only be seen in southern Australia and parts of New Zealand, whereas the northern lights cover far more inhabited areas.

Why does it always rain after a rocket launch?

A rocket launch is one of the greatest testaments to humankind's scientific achievements. There remains, however, a small mystery. Rockets always take off in clear weather, but about an hour after they launch it begins to rain. This rain is caused by the burning of the rocket's fuel.

3 . . . 2 . . . 1 . . . Launch

In order to reach space, rockets need to get out of the Earth's atmosphere, and escaping gravity is a difficult task that requires a lot of energy. Burning gasoline isn't going to get you there, so instead rockets burn a mixture of hydrogen and oxygen, which produces a much larger amount of power. When hydrogen and oxygen are burned together they form water vapor, which is released in huge quantities (this is the white stuff that comes from the boosters of rockets; it's more steam than smoke). All of this water vapor coalesces into clouds, which then produce rain about an hour after the launch.

The Difficulty of Rain

Water on its own can't make rain. Water molecules' ability to stick together is not great, and in a pure atmosphere it'd never rain. Rain droplets can only form when there is some dust or other particles in the air to which the condensing water can stick. As the water droplet begins to form around this central particle (known as a condensation nucleus), it becomes easier for other water molecules to stick to it, allowing it to condense fully. Lots of these droplets make up clouds, which eventually accumulate so much water that they become too heavy to remain floating in the air and begin to fall as rain.

Could a butterfly really cause a tornado?

There is a saying that a butterfly flapping its wings in Brazil could, in time, cause a tornado in Texas. But is this really true? Well, yes and no. But mostly no.

Chaos Theory

Edward Lorenz was a pioneer of chaos theory. He was working on statistical modeling in weather forecasting and he struggled to make the numbers work. He found that by making tiny, seemingly inconsequential changes to the initial conditions in his models, the outcomes would vary enormously. His exploration of this idea eventually became the butterfly effect.

We often think of the world as simple cause and effect, and that we can predict accurately, often with the help of computers, exactly what's going on. Chaos theory adds more complexity; it is present in any system where small changes make a big difference. The weather is only one such system, though it explains why the weather reporter is so often wrong.

Chaotic behavior is very common in the physical world and touches aspects of physics, chemistry, biology, and mathematics. It is possible that, in time, with increasing computing power, chaos theory will simply be very difficult math—but for now at least it means that so many things in our universe are essentially unpredictable.

The Butterfly's Tornado

The butterfly effect theory proposes that it is possible that the changes in air pressure caused by a butterfly's wings flapping could be the start of a long and complicated chain of reactions that eventually forms a tornado. However, the chain would be so long and complex, with so many other influencing factors, it could hardly be attributed to just the flapping of some wings.

WEATHER

Are you a meteorology buff? Test yourself with this quick quiz.

Questions

1. In what temperature is snow most likely?
2. What does light reflect through to make a rainbow?
3. Which way do hurricanes spin in the Southern Hemisphere?
4. How high are most fire whirls?
5. If you hear thunder nine seconds after seeing lightning, how far away is it?
6. What element gives the northern lights their green color?
7. What do rockets burn for fuel?
8. What is the name of the scientific theory behind the butterfly effect?
9. What is a storm surge?
10. What is the scientific name for the southern lights?

Turn to page 213 for the answers.

CAN I MAKE A DIAMOND?
WHY DOES METAL CONDUCT ELECTRICITY, BUT WOOD DOESN'T?
WHY DO THINGS MELT AT DIFFERENT TEMPERATURES?
110
100
90
80
70
60
50
40
30
20
10
0
10
20
30
40

WHAT MAKES NUCLEAR WASTE UNSAFE FOR MILLENNIA?

MATERIALS

WHY DO DOOR HANDLES KEEP GIVING ME ELECTRIC SHOCKS?

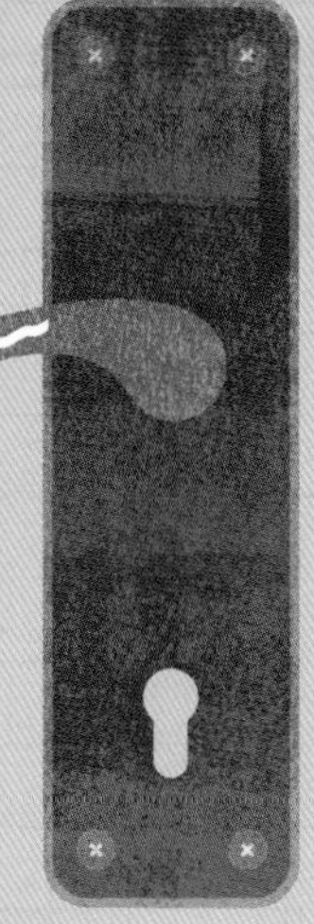

Why do things melt at different temperatures?

You can melt just about anything. But the temperature at which stuff melts can differ wildly, which is why a tub of ice cream left out on the kitchen counter will turn into liquid but the spoon next to it won't. This is because melting happens when a material has been heated up enough for the atomic bonds that keep it solid to be broken. Materials have different types and strengths of bonds so different temperatures are needed to break them.

Stronger and Weaker Bonds

There are many different types of bonds that can occur. Some rely on atoms or molecules being polar, which makes them act almost like tiny magnets, which can then attract each other. Others form when atoms are able to share electrons between themselves or where the act of losing or gaining the electrons causes them to become oppositely charged (like opposing ends of a magnet), which then causes bonds to be created. Of course, as well as the type of bonds, which atoms are forming them makes a big difference. Larger, denser atoms often make weaker bonds but may also be able to support more than one type of bond. Atoms bond to form molecules, and these molecules can also bond to other molecules, creating more bonds that will need to be broken in order to melt the material.

EXTREME BONDS

The highest known melting point belongs to tantalum hafnium carbide (Ta_4HfC_5), which melts at a scorching 6,192°F, whereas the lowest melting point is helium at -458°F, which is only one degree or so above absolute zero.

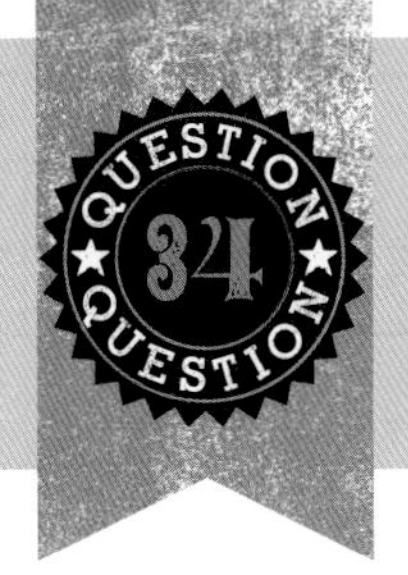

Why does metal conduct electricity, but wood doesn't?

If you've ever made an electrical circuit or even just played around with electrical components, you will have noticed that some materials are able to conduct electricity, but others, such as wood, can't. The difference comes down to the material's electrical resistance and the amount of free electrons available.

Electrical Resistance

Electricity is a flow of electrons through some kind of system, like a wire, but electricity is always hampered by resistance. Imagine electrons passing through an object as a large group of runners traveling through a dense forest. Different materials will have higher or lower resistances, which is like the trees being more abundant and closer together, meaning that the runners have to slow down or may even crash, making it harder for them to pass through quickly.

Free Electrons

Metals undergo a special type of bonding known as metallic bonding (it's what makes a metal a metal). One of the results of this is that all the outer electrons around each atom essentially become bonded to the material as a whole, rather than any individual atom. This means that when you pass a voltage through a metal, the outer electrons from the metal itself are able to flow, making it easier to carry a current.

Different Metals, Different Flow

Not all metals conduct equally. Because of the variation in their internal structure, the resistance can be increased and different metallic elements have different numbers of outer electrons. All metals can conduct electricity but some, such as gold and copper, are better than others, such as aluminum and titanium.

What makes nuclear waste unsafe for millennia?

Nuclear fission can produce huge amounts of energy without using much fuel, so it might seem like the solution to the energy crisis. The problem, however, is that the waste created is dangerously radioactive and will remain so for thousands of years because of its radioactive half-life.

Half-Life

If you were to take a lump of radioactive material with, say, 2,000 radioactive uranium atoms in it, over time they would spontaneously emit some form of radioactivity (see page 206) and decay, becoming inert. The reason this happens is due to a very complicated quantum effect, but it leaves us with an interesting pattern.

Over a set period of time, known as a half-life, roughly half of the atoms in a material will decay. In our uranium example, there will be only 1,000 radioactive atoms left. When the same time period passes again, half of the remaining atoms will decay, leaving only 500. Then again after another half-life there will be just 250, and so on and so on. It can be difficult to wrap your head around why it happens, but it's like flipping a coin for every atom each half-life. There's a 50/50 chance that any individual atom will decay

each time, so roughly half of them do. This half-life effect means that while the radioactivity of nuclear material will weaken over time, it will take a long time for it to decay to the point where it is safe for humans to be around.

Nuclear Waste

Different nuclear materials decay at a different rate. Some that are used for medical purposes have a half-life of only a few minutes, but those that come out of nuclear power plants tend to last much longer. Uranium fission produces elements such as cesium-137 and strontium-90, which have half-lives of about 30 years, and plutonium fission can produce plutonium-239, which has a half-life of 24,000 years!

RAY CATS

One of the problems with nuclear waste is keeping it safe—not just now but in the future. Much of it is stored in vast underground vaults that are sealed off, but in 10,000 years any knowledge of these places may be lost, all written and spoken languages may be dead, and even the idea of what radiation is may be gone. So how can we communicate to future generations that these are dangerous places to be avoided? One of the most interesting ideas proposed is to genetically engineer "ray cats," which would be the same as normal cats except that they would change color or glow where there is radiation present, and to then create a legend that if the cats change color, you should leave immediately. That way, even if the words we use change, as long as the knowledge is passed down that if the cats glow, you should stay away, the future people of Earth could stay safe.

Why do door handles keep giving me electric shocks?

It's a common problem. You reach for a door handle to open the door and suddenly you get a sharp, painful electric shock. Maybe it's a stair rail or a comb instead, but the point is you keep getting shocked. These things can give you electric shocks because of a buildup of static.

The Static Effect

Electricity is simply a flow of electrons from one place to another. In circuits a battery is used to push the electrons around a system, but electricity can happen naturally in the form of electrical shocks. Shocks start with static—the buildup of an electrical charge in a place. This can happen when there are too many or too few electrons. The buildup of electrons causes the object to become charged with electrical energy known as static.

This static energy can then be discharged when it comes into contact with something else. It may either absorb lots of electrons or release them very quickly, and when it does so this causes an electric shock.

Simply Shocking

You are constantly having tiny unnoticeable shocks as electrons move between you and the environment, so why can you sometimes really feel them? To reach a point where you can feel it, you have to build up a lot of static. One of the most common ways static is made is when things rub together, because electrons can be knocked off one of the materials. We can generate static in our bodies when we move around in our clothes or scuff our feet on carpets. Eventually, when we touch something made of metal—which is able to move electrons around very quickly—this can result in a painful electric shock. We often get shocks from door handles because the rubber soles of shoes keep electrons from moving between us and the ground, so it's only when we touch the metal with our hand that the shock can take place. If you find yourself getting lots of shocks, you might want to avoid wearing clothing made of materials (like wool) that produce a lot of static.

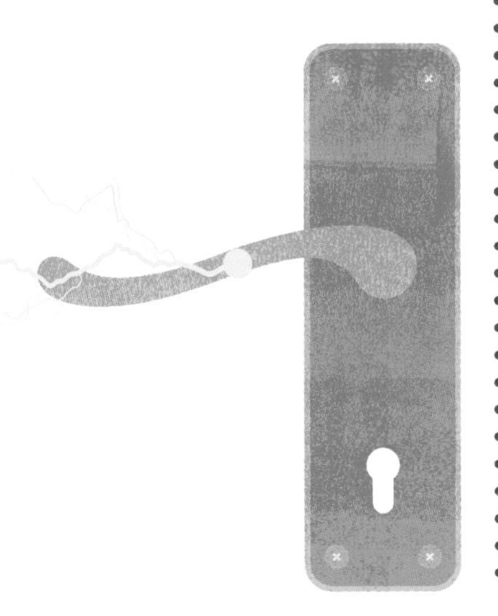

HAIR RAISING

You may have put your hands on a Van de Graaff generator before. When you do this, your hair starts to stand up on end. This is because the bell becomes positively charged as an internal belt rubs the metal and strips away the electrons. When you put your hands on it, your electrons flow to take their place and your body also becomes positively charged. This means that each of your strands of hair becomes charged with static electricity. All of this static is positively charged, and so each hair pushes away all the other hairs until they're all pushed out and floating—just like when you try to push two alike poles of a magnet together.

What can you use to float a train?

You might think that a floating train is some crazy science fiction idea, but it's already happening here on Earth today! Various floating trains have been built in Germany, South Korea, and Japan since the 1960s. These trains are able to float using the repelling power of magnets.

Older Trains

There have been lots of maglev (magnetic levitation) trains, but they have often been short-lived, used mostly for expos and only traveling short distances. The only commercially running one today is the Incheon Airport Maglev in South Korea, which shuttles to and from the airport. The Incheon Airport Maglev and all maglev trains invented before it float using electromagnetic suspension. This is where the bottom of the train is wrapped around a metal track. The train contains carefully controlled electromagnets that allow it to produce a magnetic field with the same charge as the track, keeping it floating just above the metal track.

Why Float a Train?

Floating trains come with a number of benefits. One of the most obvious is that because the train isn't touching the track, it won't be subject to friction. This means that a floating train is able to move much faster than conventional ones. They can also be made lighter than normal trains and don't put as much strain on the tracks, bringing track maintenance costs down.

It is also very easy to make a maglev train completely electric. Maglev trains don't have wheels to turn; instead, they use their magnetic abilities for forward propulsion, and this is much easier to do with electricity, rather than a gasoline or diesel engine. This in turn is better for the environment and again keeps the costs down.

Floating Trains of the Future

Future floating trains such as the Japanese Chūō Shinkansen will instead use the properties of superconductors. Superconductors are very special materials that, when cooled down to very low temperatures of at least -328°F, will cease to have any electrical resistance at all. As a superconductor is cooled down into its superconducting state, it also takes on another property known as the Meissner effect, which causes the pinning of magnetic fields. What this means is that if a magnet is held above the superconductor as it cools, then when the process is complete it will float there for as long as the superconductor is kept cold.

Other Uses of Superconductors

While the role of a superconductor in floating-train technology is very exciting and useful, its zero-resistance quality is much more interesting. It will allow for better electronics that are faster and don't heat up, which is a major technological problem at the moment. They could allow for better power transfer with less loss, reducing the amount of power that is needed to be produced (see page 105). They also have uses in medicine, such as in MRIs, and in various scientific experiments, such as in the large hadron collider at CERN. The major problem with superconductors is that all the known ones still need to be kept very, very cold, so the race is on to find a "room temperature superconductor"; if one is ever found, it will revolutionize our world.

Can I make a diamond?

You may have been led to believe that diamonds are rare and precious. Some sort of magical stone forged only in the heart of the world. But diamonds are just specially arranged carbon atoms, and it is possible to make one in a lab.

Rearranging Carbon

In the element carbon, each atom allows four chemical bonds to it. Because there are multiple bonds available, there are lots of different ways that the atoms can be put together. Humans are largely composed of carbon, but our carbon molecules look very different from the largely unorganized arrangement within the carbon molecules in coal. Carbon can also be structured to make graphite for pencils (see page 83), and single layers of graphite are a material in their own right, graphene.

Diamonds are just another arrangement of carbon. They are made of carbon atoms arranged in a tetrahedron shape. These tetrahedrons then build onto each other, creating the very hard crystal structure we know as diamonds.

Making Diamonds

Diamonds are made naturally in the Earth's crust by the large pressure and high temperatures the carbon is exposed to. It is possible to make synthetic diamonds by re-creating similar conditions with industrial machinery, applying huge pressure to press the atoms into shape and subjecting them to scorching temperatures to burn away all but the toughest diamond bonds.

Synthetic diamonds are functionally identical to the ones found in the ground, but they are usually made to be much smaller to keep costs down. These diamonds are often used in construction tools for their hardness, and in some electronic devices. Some even make their way into jewelry.

Why do things expand when they get warm?

We know intuitively that when things get hot they swell in size. Doors might stick, pavements crack, and bridges have to be built to allow for it. But why this happens isn't immediately obvious. The reason things get bigger with heat is that they have more energy.

How Does More Energy Make It Bigger?

Temperature is really a measurement of how much the atoms in something are moving around. The hotter something is, the faster the atoms are vibrating or moving around. So as you heat something up, each atom will move more. The increased movement of atoms means that it takes up more space. While each individual atom might only take up a tiny amount more space, when this happens to all of them in an object like a rock, it will cause a noticeable amount of expansion.

Freezing Expansion

A select few materials also expand when they freeze. You may have discovered this yourself by putting a full bottle of water in the freezer and opening it up later to discover the bottle has cracked. In liquid form, all the atoms in a substance are able to slosh around freely, but when some chemicals turn into a solid, they settle into crystalline structures. These structures push individual atoms farther apart from each other and lock them in place, making the material as a whole expand.

What makes nonstick pans nonstick?

It's always annoying when you're cooking a meal and find that it's gotten stuck to the bottom of your pan. Modern culinary technology, however, has introduced nonstick pans that help to avert such kitchen disasters through the use of low-friction coatings.

A Sticky Situation

It may surprise you to learn that your rapidly burning eggs don't stick to the pan like glue. Rather, they (and any other sticking food) actually get caught in tiny holes, scratches, and other damaged areas of the pan's surface. Some of these are made naturally during manufacture, but they can also be caused by overly vigorous washing, which is why older pans tend to stick more.

Fixing It with Friction

Nonstick pans are made in the same way as normal pans but are then coated in special materials called plastic polymers. Plastic polymers are long chains of chemicals (a bit like spaghetti) that are able to wind together into large sheets. These sheets of material are very strong, they don't chemically bond easily, and they have a low friction. This all means that food will be able to glide easily over the polymer surface. Many people ask how you can stick the nonstick polymer to the pan in the first place! This is possible because the pan starts with a very rough surface, and when the plastic is sprayed on it is able to stick itself to this before it forms the smooth nonstick layer itself.

MATERIALS

Do you know what's what when it comes to what things are made of? Try this quiz to see.

Questions

1. What recently developed material has the highest known melting point?
2. What is the thing that makes a material a metal?
3. How long is the half-life of radioactive element strontium-90?
4. What is the element that forms a diamond?
5. What is the buildup of electricity in a material called?
6. What temperatures do superconductors need to be cooled to in order to work?
7. What kind of motion in particles causes expansion in many types of materials?
8. What chemical is commonly used to make nonstick pans nonstick?
9. In what kind of medical equipment might you expect to see a superconductor?
10. Where are diamonds naturally found?

Turn to page 213 for the answers.

WHY CAN YOU BURN THINGS WITH A MAGNIFYING GLASS?
HOW LONG WOULD YOU LAST IN SPACE WITH NO SPACE SUIT?
WHAT MAKES ATOMIC BOMBS EXPLODE?

HOW DO SUNGLASSES MAKE THINGS DARKER?

HOW DO PLANES STAY IN THE SKY?

TECHNOLOGY

Why is there a mesh on my microwave?

Have you ever noticed when looking into your microwave that there is a mesh on the back of the window? Microwave ovens use microwaves to heat up the water in food; if let loose, these waves can be dangerous to humans. The mesh has holes smaller than the microwaves, so it keeps them safely contained within the oven.

Lunch Problems

A peryton (named after a mythological animal) was a phenomenon first observed by the 210-foot Parkes Observatory radio telescope in Australia in 1998. These were millisecond-long bursts of short-length radio waves with no known source. It was quickly determined that they weren't coming from deep space, where the telescope was pointed, but there were a lot of theories as to the source, ranging from aircraft signals to solar flares. It took until 2015 before it was realized that the cause was impatient scientists at the observatory opening the kitchen's microwave door before it had finished and thus releasing into the air some microwaves that were usually kept in by the mesh.

A Chocolaty Measurement

Using just a microwave and a chocolate bar, it's possible to measure the speed of light! Place the chocolate on a microwaveable plate, make sure the chocolate bar will not rotate by removing the microwave plate, then set it going until it starts to melt in a few places (about 20 seconds). The waveform has a peak and a trough, each of which produce a melted spot in the chocolate. You need to measure from peak to peak or trough to trough to get the wavelength of the microwaves, so the easiest way is to measure from one to the other then double it because the distance will be the same. Multiply this by the frequency of your microwave oven (standard microwaves are about 2.49×10^9 hertz) and this should give you the speed of light. The real speed of light is 983,571,056 feet per second—how does yours compare?

How long would you last in space with no space suit?

Space is a very expensive place to go. Part of that cost is the multimillion-dollar space suits that astronauts wear. So in order to cut down on expenses, why not go without? Well, trying to go into space without a space suit would mean a very quick and painful death for four main reasons.

Oxygen

In space there is no air; there's not much of anything at all, really. Space suits provide the constant flow of oxygen needed to keep us alive. An average person can hold their breath for a couple of minutes, but the conditions of space would draw the air out of your lungs, cutting this down to about 15 seconds before you lost consciousness.

Temperature

Space is very, very cold. At -455°F, it's only a few degrees above absolute zero. These temperatures would cause every part of your body to freeze solid and crack. This wouldn't happen immediately, as your normal body temperature would have to cool off, which might take a minute or so.

Pressure

There's not a lot of stuff in space, but there's plenty in you. This difference in pressure would cause a huge number of problems, as the stuff inside you would naturally try to push outward. Some of the materials in your blood would turn back into gas, creating bubbles; these would then cause your body to inflate. The pressure would also cause any liquid on the surface, such as around the eyes or in the mouth, to start to boil despite the low temperatures (see page 150).

Radiation

Even if you somehow managed to survive all of the other problems, being in space exposes you to a huge amount of radiation. This radiation comes from stars and other sources, and the high levels of UV rays, X-rays, and potentially gamma rays would damage you irreversibly on a molecular level.

How does a freezer stay cold?

One of the greater unsung innovations in the modern world is the ability to keep things (especially food) cold. It allows for movement of fresh food over huge distances and longer-lasting stock, but how does it work? A freezer passes a cold fluid through the inside, constantly pulling the heat out.

Heat Transfer

Heat flows from a hot place to a cold place. If you take two metal blocks of different temperatures and make them touch, the heat from the warmer one will flow into the colder one until they both reach the same temperature (see page 135). If you look on the inside of your freezer, you will likely notice some kind of snaking tube. A very cold liquid passes through this tube and the heat inside the freezer flows into the liquid, which then takes it out of the device.

Keeping It Cold

Taking the heat out of the system is easy, but it is then in the liquid. If left in the liquid, eventually everything would warm up to room temperature, so the heat needs to be removed from the liquid. This is done by taking the warmed liquid and converting it into a high-pressure vapor, which is then condensed into a high-pressure liquid (this is done in the grill on the back of your freezer). The liquid is then put into a capillary tube, which is a tight coil of tubing (it looks almost like a spring). The shape of the tube means that there is a huge drop in pressure between the top and the bottom. This pressure drop causes the liquid to cool, allowing it to flow back into the freezer and remove more heat, keeping the whole system cool.

How does a pencil make a mark?

If you need to write something down, you might use a pencil. But have you ever questioned how a pencil makes a mark? Pencils take advantage of the special composition of graphite to leave thin layers behind on the page.

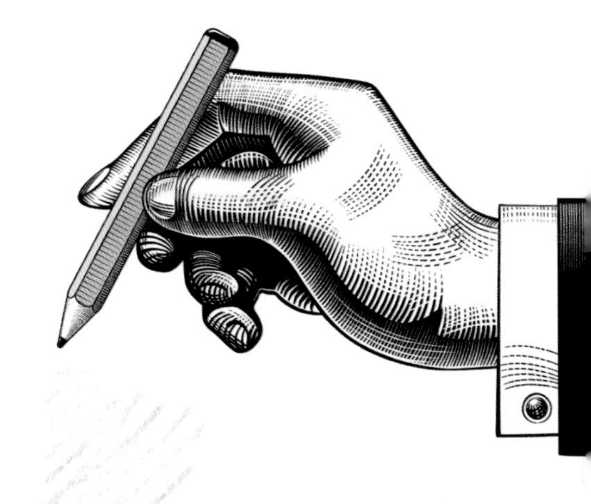

Layers Like an Onion

Pencils are made from graphite. Graphite is a form of carbon that is structured in a very special way. The carbon atoms form a regular pattern (usually hexagonal) in a large sheet. Multiple sheets are then able to build up on top of each other in a layered structure. The bonds in the sheets are very strong and difficult to break, but the bonds between the layers are very weak. This means that they are able to slide off with only a small amount of force. A pencil has a very thin core of graphite, known as a lead. When you use a pencil, you are applying pressure to the graphite, which causes some of its layers to slide off and remain behind on the page.

Space Pencils

You may have heard the old tale. When going into space, the United States invested millions of dollars and years of research into inventing a pen that would work in space. One that could work upside down, underwater, in zero gravity, and in a wide range of temperatures. The Russians used a pencil. This is amusing but untrue; both the U.S. and the USSR originally used pencils but quickly switched to pens. This is because, being made of graphite, pencil leads are easy to break and small flecks could come off. Graphite is conductive, so if a free-floating piece of graphite made it into some electrics, it could cause all sorts of havoc.

What makes atomic bombs explode?

Atomic bombs are among the most destructive things humans have ever made, with the power to level whole cities and cause devastation for miles around. Their massive energy is released by the reactions of individual atoms, which can be done in one of two ways.

Fission Bombs

Fission bombs are the kind that were dropped on Japan during World War II. They are made up of large elements that contain lots of neutrons and protons in their nucleus, such as uranium or plutonium. A neutron is fired at the nucleus of a large plutonium atom, which causes it to split apart, releasing some energy and some neutrons. These neutrons then collide with other plutonium atoms, causing further reactions and releasing more energy and yet more neutrons which do the same, and so on, in a chain reaction that causes the release of an enormous amount of energy. A single pound of plutonium is equivalent to about 10,000 tons of TNT.

Fusion Bombs

Fusion bombs, often known as hydrogen bombs, or H-bombs, are much more advanced and destructive than fission bombs. They use X-rays or gamma rays to compress the hydrogen, heating it up massively. This causes the hydrogen to go thermonuclear and release huge amounts of energy and very high-speed neutrons, which then cause fission in materials elsewhere in the bomb. About half of the energy released comes from the initial fusion process and the rest from the additional fissile materials. The largest fusion bomb ever exploded was the Tsar Bomba in 1961, which exploded on Severny Island north of the Russian mainland with a force equivalent to 55,000,000 tons of TNT.

Why are there so many kinds of light bulbs?

Buying light bulbs can be a bit of a chore; it's very easy to get the wrong kind accidentally. Aside from the different sizes and fittings, there are different types of bulbs, too. Different bulbs use different methods to produce varying amounts of light with varying efficiencies.

Incandescent

This is the traditional type of bulb. The electric current is passed through a tightly wound coil wire (usually made of tungsten) called a filament. The extreme heat that this causes makes the filament glow. Incandescent bulbs produce a bright warm light but are incredibly inefficient, turning only about 2 percent of the energy into light and much of the rest into heat.

Halogen

Halogen bulbs use the same coil as in incandescent bulbs, but the rest of the bulb is filled with gases including a halogen (such as bromine or iodine). These gases are able to capture some of the material that is burned off the coil initially. This allows it to be

deposited back onto the filament, thus recycling the material within the bulb, allowing it to work for longer.

Fluorescent

The long white bulbs that you may have seen in schools or offices are fluorescent tubes. Within the bulb is a mercury vapor; electric current is passed through this vapor, causing it to excite. This causes the vapor to release electromagnetic waves, which then cause a phosphor coating on the inside of the bulb to glow, producing the light. This is far more energy-efficient than incandescent bulbs.

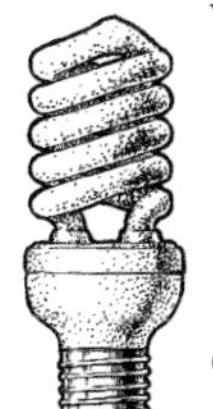

LED

An LED (light-emitting diode) bulb is made up of special tiny electrical components. When supplied with a current, they use quantum effects to release light. LEDs use a very small amount of power and are far more efficient than other types of bulbs. Thanks to recent improvements, LEDs can now be as bright as incandescent bulbs.

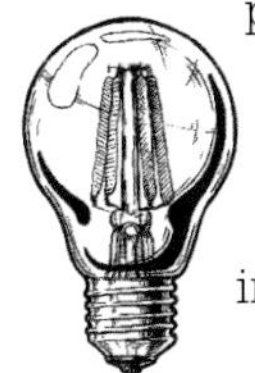

What makes a crystal watch tick?

Watches have been used for over a hundred years to tell the time. But fitting the physics of a grandfather clock into a watch requires a special material. In a crystal watch, a tiny quartz crystal vibrates in a regular pattern, which forms the basis for its timekeeping.

Keeping Time

All analog clocks work on the same principle: Some sort of mechanism moves with a regular period. Grandfather clocks use a pendulum that takes the same amount of time to swing to and fro. Alarm clocks and wind-up watches use a spring balance wheel, which also moves back and forth. Even modern atomic clocks use the movement of an atom to keep a regular time interval.

Crystal Power

Crystal watches (often called quartz watches) contain a small, roughly cylindrical piece of silicon dioxide crystal. Silicon dioxide has a property that makes it piezoelectric. This means that when the crystal is squashed or stretched, it creates a small electrical current. Crucially, this also works the other way—applying electricity to the crystal causes it to squash and stretch repeatedly, making it vibrate. Quartz watches are often crafted so that the crystal vibrates 32,768 times per second. The electronics or gears in a watch convert this into one movement every second. This drives the second hand, which is hooked up to the other parts of the watch, making them all work.

> "When the crystal is squashed or stretched, it creates a small electrical current."

Exact Time

Knowing the exact time can be very important and has been for many years. Especially when traveling at sea, good timekeeping can be the difference between life and death. Before the creation of high-quality clockwork in the mid-eighteenth century, knowing the exact time on a ship could be difficult—clocks with pendulums would be affected by the shifting of the boat in the sea. Many solutions to this problem were proposed, some more practicable than others. One man suggested giving each boat a dog that had been cut with a particular knife; this same knife would be plunged into a mix of the dog's blood and some chemicals by someone on land who had access to a reliable clock at exactly noon each day. The idea was that the dogs would all yelp at exactly the same time, so the ships' crews would know what time it was. Fortunately, this nonsensical idea was almost immediately thrown out.

Today the exact time is kept by over 400 atomic clocks in labs across the world. These use the vibrations of a cesium atom to keep an exact measurement of the passage of time and are regularly compared to keep a single standard time. Atomic time (TAI) is based on the precession of the Earth around the Sun, whereas standard time (UTC) is based on the amount of seconds that have passed since we first started counting. Interestingly, because the Earth does not move perfectly around the Sun, the international standard atomic time is 37 seconds ahead of the standard time we use in our day-to-day lives.

Can you really unlock a car using your head?

You may have heard the old wives' tale that by holding your remote car keys to your head you can extend their range, letting you unlock your car from farther away. Incredibly, this is actually true—your body can act as an amplifier.

Mini Transmitters

Remote car keys are short-range radio transmitters; they produce a radio wave with a specific coded signal that is received by the car, causing it to unlock. Radio waves are created by the movement of charged particles like electrons. How the charged particles move changes how big the waves are and how far apart the peaks of waves are. Car keys produce a frequency of 315 MHz if made in North America and 433.92 MHz if made elsewhere, and typically have a range of 15–65 feet. Beyond that range, the signal sent by the key will become too weak and unreadable by the car, meaning it won't unlock.

Body Amplifier

A surprisingly large proportion of your body—especially your head—is made of water. When you put radio waves through water, it can cause the water molecules to move in phase with the electromagnetic effect of that wave, which then causes them to mimic the signal. This mimicking will then cause the radio signal to become more powerful. So when you hold your remote car keys up to your head, their range is thus extended, which allows you to unlock your car from farther away.

How do planes stay in the sky?

It likely hasn't escaped your notice that airplanes are very big and very heavy. Despite this, they manage to fly around as though there is no gravity pulling them down. Airplanes can stay in the sky because they are able to produce enough upward force (called lift) to counteract gravity.

Generating Lift

To put it simply, in order to generate lift, the wing is designed so that there is more air hitting the underside, and with more force, than is hitting the top. Airplane wings are shaped so that they are flatter on the bottom and are angled slightly upward. This means when the plane is moving forward the bottom surface of the wing bumps into lots of air molecules. This bumping produces upward force. As the top is more curved and sloped back, far fewer molecules are able to bump into it and push it down, producing an overall upward force on the wings.

Getting Higher and Lower

In order to take off, a plane needs to get higher and so it angles upward. This means that the bottom of the wing is hit by more molecules so it produces even more lift, which causes the plane to ascend. Conversely, in order to descend, the plane tilts forward so that there is less lift produced by the wing. Plane wings are carefully designed so that when they are flying perfectly level, the forces of gravity and lift are exactly canceled out, so it will remain at the same altitude.

How do sunglasses make things darker?

If it's sunny out, you might throw on a pair of shades. These marvelous inventions are able to make even the brightest sunlight bearable. High-quality sunglasses make things darker by polarizing the light, which means reducing the amount that reaches your eyes by only allowing certain phases of light through.

It's All About Rotation

Light is an electromagnetic wave; this means it acts as a flat, two-dimensional curve. Light also has a property called phase. Phase is the angle of rotation of the wave of light. A normal light source will produce lots of light and even the light it sends in the same direction can have a different phase, meaning the angle of the light is rotated.

A Protective Grid

The lenses in sunglasses are made of a polarizing filter; this is a series of tiny bars (like a prison cell) that are too small to be seen with the naked eye. What this does is to filter out certain phases of light. The light that is in phase with the bars is able to slip through, but if it's slightly rotated, then it'll be blocked from getting through. Everything appears darker when you put on sunglasses because the amount of light that is able to pass through the filter is only a fraction of the original light. Sunglasses can be made to be darker or lighter by reducing or increasing the size of the gaps in between the bars.

Why can you burn things with a magnifying glass?

It's something many people have done as a child—used a magnifying glass to burn a piece of paper (or ants, if you're cruel). But how can a simple bit of glass manage to set something on fire? It happens because magnifying glasses are able to focus the Sun's rays onto a smaller point.

Light Manipulators

One of the easiest ways to manipulate light is through the use of lenses. These are curved pieces of glass or other transparent material. The shape of the piece affects light in different ways. Convex lenses are curved outward, and this causes light passing through the lens to bend inward. Concave lenses, on the other hand, are curved inward and cause light to bend outward. By varying the curvature of the lens, it is possible to manipulate the size, shape, and direction of light very accurately. A magnifying glass is one big convex lens, so it directs lots of light onto a very small point. With more of the Sun's rays focused onto a smaller point, they are able to heat it up and burn whatever they are focused on.

Using Lenses

It's not just magnifying glasses that use lenses. Eyeglasses also contain lenses, which are able to change the light entering the eye, making it easier for the wearer to see. Telescopes and microscopes are also made of a series of lenses that either make the incoming light more compact, so that the image is more zoomed in, or spread the light out, so that the image is larger. Your eyes also have lenses in them, which is why you should never shine a laser pointer at your eye—the lens will focus the beam to a very powerful point on your retina, possibly blinding you.

Where does a compass point?

Maybe you have used a compass before; they're a staple of orienteering activities. Everybody knows that the needle of a compass always points toward the North Pole. Except that it's actually pointing to the South Pole!

Which Pole Is Which?

This may be something of a trick question, because compasses do point toward the north, but that's not the same thing as the North Pole. The Earth actually has a few different poles. The one you're likely thinking of is the Earth's geographical pole, which is where the Earth's axis of rotation intersects the surface with (as shown on maps) north at the top and south at the bottom.

Magnetic Poles

But the Earth's magnetic poles are a bit different. For a start, magnetic north moves around a bit—at the time of writing, the magnetic north pole is nearly 435 miles away from the geographic North Pole and can move at speeds greater than 30 miles per year. The magnetic north and south poles also don't line up; if you were to draw a line right through the Earth from the magnetic north pole it wouldn't come out through the magnetic south pole. These things are due to the processes within the Earth's core and solar winds.

Even more confusing is that what we call Earth's magnetic north pole is actually its magnetic south pole! In magnets, the field always flows from north to south. On the Earth the field flows from the magnetic south pole over to the magnetic north, pointing your magnetic compass needle as it does so. So this means your compass is actually always facing the south pole of the Earth's magnetic field.

> "Compasses do point toward the north, but that's not the same thing as the North Pole."

TECHNOLOGY

Got every gadget going and know how they all work? Show off your knowledge with this quiz.

Questions

1. What delicious treat can be used to calculate the speed of light?
2. What material is used to make a quartz watch work?
3. What temperature is space?
4. What property can be changed in a liquid to cool it down?
5. What are pencil leads made from?
6. What are the two types of atomic bombs?
7. Is there more air pressure on the top or bottom of an airplane wing?
8. What in your head lets you transmit electromagnetic signals?
9. What is the most efficient type of light bulb?
10. What effect can cut down the amount of light that reaches your eyes?

Turn to page 214 for the answers.

HOW DOES WI-FI WORK?

WHAT MAKES A QUANTUM COMPUTER SPECIAL?

WHAT'S THE DIFFERENCE BETWEEN A HOUSEHOLD BATTERY AND A CAR BATTERY?

HOW IS A POWER PLANT LIKE A KETTLE?

COMPUTERS AND ELECTRONICS

How do computers store 1s and 0s?

We use computers for nearly everything without ever really knowing how they work, or what the physical components look like. While we generally understand that computers run off 1s and 0s, how this actually happens is a complete mystery to many. Computers physically store 1s and 0s inside magnetic domains.

Magnetic Bits

Computer hard drives are made out of disks that are able to rotate, and the information stored on them can be read or written using a head in a process that works like an old record player. In record players, the information is stored in the grooves in the record that the needle "reads" as it passes over, but in hard drives it's encoded in tiny sections of the disk called magnetic bits. (A byte is made up of eight bits.) These magnetic bits are made out of several even smaller "grains," which are naturally occurring structures in metals. These grains act like tiny magnets, and a magnetic bit can be made by linking some of these grains together to act as one bigger magnet. It is possible to store data using these magnetic bits. An individual magnetic bit can be pointing left or right (or up and down, depending on how it's built). When you want to access the information, a read head is passed over the top of the bits. When it detects that there has been a change of direction from one domain to the next, it reads that as a 1. If there has been no change, it instead reads a 0. A hard drive can either read or write (by flipping the magnets) millions of bits every second.

“Computers physically store 1s and 0s inside magnetic domains.”

The Future of Storage

There is a constant drive forward for more storage on hard drives and for them to become even smaller as technology changes. One of the types of hard drive just on the horizon is heat-assisted magnetic recording (HAMR). An obvious way to make a hard drive fit more data onto it is to make the bits and grains smaller. However, when you make them smaller they become more susceptible to environmental influences that can cause the magnetic direction of the bit to change randomly, which would corrupt the data. To combat this, a magnetic layer with a higher coercivity can be used; this means that the magnetic directions are more strongly held in place. However, when the higher-coercivity material is used, it also makes it harder to write new data onto the hard drive, which slows down the computer. In HAMR, before trying to write new data, a laser is used to heat up the material, which lowers its coercivity. The information is then written onto the hard drive, which then cools, locking it into place. This means that more data can be stored in the same space.

SOLID STATE DRIVES

Instead of magnetic materials, solid state drives (SSDs) use electrical transistors that are either open, allowing electrons to flow through (a 1), or off, stopping them (a 0). This allows for the information to be recorded and written faster and makes the device more durable, as there are no mechanical parts to break down. However, SSDs are more expensive and store less data than traditional hard drives.

How does Wi-Fi work?

Wi-Fi (which is a marketing term that doesn't stand for anything) is absolutely everywhere these days; homes, offices, cafés, and even some entire cities have Wi-Fi. All Wi-Fi works by transmitting information-carrying radio signals between an internet-connected hub and the Wi-Fi enabled device.

Creating Radio Waves

Before we look at what makes Wi-Fi waves special, we need to consider how radio waves work normally. Radio waves are very easy to make; you just need a simple antenna. Radio waves are a type of electromagnetic radiation (just like light) and they can be created by simply moving some kind of electrically charged particle, like an electron. In a radio antenna, electrons move up and down a metal pole to create radio waves. By moving them up and down at different speeds, radio waves of different frequencies can be made. Anything that is powered by electricity, which involves the movement of electrons, also gives off radio waves; since most modern technology is powered by electricity, there are lots of radio waves around you all the time. Radio antennae, however, are specially designed to generate a single, large, and clear radio wave that will be able to travel long distances.

IS WI-FI DANGEROUS?

Some people are concerned about the possible health risks associated with living in an environment that is constantly full of radio waves. However, on this topic the science is conclusive. There is no danger from the radio signals put out by Wi-Fi devices, mobile phones, telephone poles, or any other modern electronic equipment. This is not only because the radiation from them is non-ionizing and thus safe but also because any amount we produce through our technology is massively dwarfed by the colossal amount of radiation coming from space that passes through us without harm every day.

Carrying Information

We often picture radio waves as simple regular patterns, but in order to be useful they must be able to carry information somehow. But how? To transmit information, the normal radio wave frequency (for Wi-Fi this is usually 2.4 GHz) is "enveloped" by an information signal. This means that the size of the wave is squished into the shape of the information signal before being broadcast out. This is like taking a stiff wire with just a little bit of give in it and bending it into a shape; the main curve is still the same but the finer details contain the information. When the signal then reaches a device such as a phone or a laptop, the information can be extracted from the shape of the radio signal. This allows for rapid data transfer, which makes wireless internet usage possible.

What's the difference between a household battery and a car battery?

Batteries save us all from a cord-filled existence, and they come in all shapes and sizes. But if they all perform the same role of producing power, why can't we use any battery for every purpose? Why not power your car with AAs? The reason all batteries aren't interchangeable is that they produce different amounts of current.

Voltage and Current

Voltage is a measurement of electrical potential energy. Imagine climbing up a ladder; as you climb, the distance between you and the ground gets greater, leading to more gravitational potential energy. You don't notice the energy as you climb, but if you fell, you would. In batteries, this potential energy is made by a physical gap in the circuit; charge needs to build up on either side, until it becomes great enough that it overcomes the barrier and electricity is able to flow. The voltage of a battery is the potential energy needed to overcome the gap. Current, on the other hand, is a measurement of how much electricity is flowing around a circuit. The more electrons or the faster they are moving, the higher the current.

Different Batteries for Different Jobs

It is possible to buy both 12-volt household batteries and 12-volt car batteries, but you can't use one for the purpose of the other. This is because they need to do different jobs, so they provide different things. A household battery for a remote or something similar might need to provide a small but constant current, whereas a car battery will instead produce a short but large current in order to start an engine. Household batteries normally only produce a constant current of around 0.05 amps, whereas a car battery may provide a short burst of upward of 400 amps!

Current Kills

Voltage is not that dangerous. If you've ever touched a Van de Graaff generator (see page 71) in a science class or at a science fair, then you will have received a shock in excess of 100,000 volts. Perhaps a little painful, but not unsafe. Current, on the other hand, can prove deadly. Electrical currents passing through the human body can mess with all sorts of normal body signals, causing anything from muscle spasms to fully stopping the heart. The heat resulting from the body's natural electrical resistance can also cause major internal and external burns. Anything over about 0.1 amps can be fatal.

ARE CAR BATTERIES SAFE?

If normal car batteries are able to produce over 4,000 times the amount of current needed to kill you, does that mean they are incredibly dangerous to us humans? The answer, thankfully, is no. The key to this is our bodies' natural electrical resistance. Human skin is a very poor conductor of electricity (see page 70), so in the end, while car batteries are able to produce very large currents, contact with human hands will only result in a slight tingle as opposed to a deadly shock. This is not to say that you should touch a car battery or be careless with them. Any type of battery, if damaged, old, or used improperly, could cause serious injury, especially larger ones from cars.

Why are fiber optic cables the future?

Since the days of the telegram, information has been transmitted through copper cabling. From the transatlantic cable to phone lines in homes, copper has been the material of choice. More recently, however, fiber optic cables are becoming more common as they are able to carry much more information more reliably.

Underground Copper Cables

Information signals can be sent along copper cables in the form of pulses of electricity. In the early 1800s, this was originally just simple on or off signals used for Morse code in telegrams, where a combination of long or short pulses can be converted into letters. As things became more advanced, it was possible to send more than a single signal down the same wire by varying the intensity and phase of the current. These copper cables were often bundled together in some protective shielding and would be laid down underground, where they would be kept safe from accidental damage and external forms of interference.

What Are Fiber Optic Cables?

You may well have come across fiber optic cables before: They are a staple of stadium flashing wands and ultra-modern lighting. They consist of a very simple glass or transparent plastic tube through which light can be transmitted from one end to the other, even as they move and bend. A strand of optical fibers allows light to move down the tube without leaking out. This is due to total internal reflection—the nature of the material used for the optical fibers means that any light that hits the side of the tube bounces off and back into the fiber. A dark cladding is put around the glass to help with this process.

Doing a Better Job

Fiber optic cables work using pulses of light. Like the original copper cables, they send 1s and 0s as on or off signals from one end to the other. The difference, however, is that it is possible to vary the frequency of light being sent to allow for at least ten times the number of signals as even the most advanced copper cabling. It's not just information carrying where fiber optics win out. They have better transmission length, meaning that they are able to carry information over a much longer distance without the data starting to degrade. They are much more durable than their copper counterparts, as well as longer lasting. Even protected copper cabling can oxidize (meaning that exposure to air changes the material) over time, which causes it to be less efficient. Furthermore, as copper cables use electricity, it is possible for them to be affected by external fields that can damage their signal, whereas fiber optic cables avoid this problem by using light. Fiber optic cables are also more secure: It is impossible to tap fiber optic cables and they don't leak out signals like copper does. The only reason fiber optics aren't everywhere at the moment is that they're more expensive and existing networks of copper cables are already in the ground. However, as copper gets more pricey and fiber optics become cheaper, as the old cables need to be replaced it seems the future is in fiber.

How do speakers work?

Just one of the things taken for granted today is the easy access to speakers that can produce highly realistic sounds. Gone are the days of scratchy gramophones that trace out and amplify physically recorded waves. Modern speakers work by manipulating a magnet to move back and forth, which produces a sound.

Speaker Construction

Speakers consist of a semiflexible cone that is attached to a small electromagnet (a device that is only magnetic when a current is passed through it). A larger permanent magnet sits close by, and when a current is passed either one way or the other through the electromagnet, it is either repelled or attracted to the larger magnet. All sounds, from your voice to the roar of a waterfall, are the result of vibrations, usually in the air. A speaker uses the push and pull of the magnet on the bottom of the cone to cause air vibrations within it, which when carefully controlled can create music, speech, or any other kind of noise.

Different Speakers, Different Sounds

Any audiophile will be able to tell you about the many different types of speakers that exist, from subwoofers to "tweeters." While they all operate on the same basic principle, they do slightly different things. Generally, a variation in the shape of the cone means that different speakers are able to produce better-quality sound at higher or lower frequencies. Headphones also work in the same way as normal speakers, albeit on a much smaller scale. They compensate for this fact by only needing to vibrate the relatively small volume of air in your ear canal.

What do transformers do?

Transformers are everywhere, from the roadside building plastered in warning signs to the strange box on the chargers for all your gadgets. Despite their ubiquity, however, you may not even know what they're for. Transformers increase or reduce the voltage of electricity ready for use or transmission.

What Goes on Inside the Box

A transformer consists of three main elements: the incoming primary wire, the outgoing secondary wire, and a magnetic core. When the primary wire has a current applied to it, this induces a magnetic field in the core which then, in turn, induces an electrical current in the secondary wire. This process allows for power to be transferred from one side of the transformer to the other. What changes across the transformer is the electricity's voltage; how it changes is dependent on the number of coils of wire on either side. If the secondary wire has more coils than the primary wire, then the voltage is increased; if the reverse is true, then the voltage is decreased.

The National Grid

In some countries, power plants produce electricity at 25,000 volts (see page 106), but this needs to be changed before transmission. As electricity travels along overhead lines, the lower the current is the less power is lost, and so the voltage is increased to 400,000 volts. This voltage of electricity would blow up all the electronic devices in homes, so before it reaches them it goes to small local substations where it is reduced to 120 volts (known as household power). This power is then sent to houses and businesses where consumers can plug in devices that are designed to work at this voltage or that have their own small transformers (usually as part of the power cable) to reduce the voltage down to the correct level.

How is a power plant like a kettle?

Keeping the lights on is no easy feat. Enormous buildings are dedicated to producing enough energy for all our daily needs. They may seem incredibly advanced, but power plants have something in common with a very ordinary household object. While it's for very different reasons, power plants and kettles both produce a lot of steam.

Generators

It is possible to induce an electrical current in a wire by moving it in a magnetic field. The most efficient way to generate electricity by this method is to rotate a coil of wire within a magnetic field. Doing so at a constant rate provides a stable current of electricity passing out of the ends of the wire. As the wire will be traveling through the field one way for half a turn (up) and then the other for the second half of the turn (down), this causes the electrons in the wire to flow one way and then the other. This is known as alternating current (AC) and is the normal type of electricity you find in your home.

Giant Kettles

While it is possible to generate alternating current using a simple hand crank or wheel to turn a coil of wire, at a power plant huge amounts of electricity must be generated, so a much more efficient way of turning a coil of wire is needed. Almost all power plants do this in the same way: Some fuel is burned and the resulting heat is used to boil a vast vat of water like a kettle. Oil, gas, coal, and even nuclear power plants all use their respective fuels to heat up the water. Once the water reaches boiling point, a super-hot steam (up to 1,100°F whereas a standard kettle only produces steam at 212°F) rises rapidly and passes through a series of turbines, causing them to spin. These spinning turbines are hooked up to generators that are used to produce the electricity that is sent across the country.

Renewable Energy

Most renewable forms of energy also rely on the rotation of a turbine to generate electricity. In a wind farm, it is obviously the wind driving the turbine, while in hydroelectric power plants the turbines are turned by large volumes of water falling from a dam. Even in plants that use biomass and geothermal energy, that energy is used to boil water for the steam that can then turn the turbine. The exception is solar power. There are some solar power plants that use mirrors to reflect the Sun's rays into the plant to heat water and create steam. However, it is more common that the photovoltaic cells in solar panels absorb sunlight and are then able to release electrons, which creates the desired flow of electricity.

FUSION: FUEL OF THE FUTURE!

Scientists have billed fusion energy as a promising and bountiful energy source in the next few decades. Fusion replicates the conditions of the Sun's core and is able to force two hydrogen atoms to fuse together into a single helium atom, a process that releases 10,000,000 times as much energy per pound than any fossil fuel. But even when fueled by this advanced form of energy, power plants will still use the energy to heat up water, to create steam to turn a turbine.

What makes a quantum computer special?

Computers are always getting smaller and faster, but we're starting to reach the limit. Transistors (the smallest part in a computer) are already about 500 times smaller than a red blood cell; at this level, quantum effects mean that traditional components just stop working. What makes quantum computers special is that they take advantage of quantum effects to store and process huge amounts of information.

Qubits

Information in normal computers is stored as a bit, as part of a magnetic medium (see page 96). A bit can have two values, either 1 or 0. A qubit, on the other hand, is any property of a material that can only have two possible states, such as the spin of an electron (up and down) or phase of a light photon (horizontal and vertical). The qubit can use the quantum effect of superposition to be a bit of both at the same time, and it only becomes one or the other when you use it. This means that you can store multiple pieces of information in the same space. Old 8-bit video games used 8 bits to store a single piece of information, but if they were to use 8 qubits they could store 256 instead because the qubits can be all of the possibilities of 1s and 0s at the same time. This number grows exponentially, too: Many modern computers use 64 bits to store information, meaning that an equivalent qubit computer could store 18,446,744,073,709,551,616 times the information!

Why Should I Care?

While you shouldn't expect to see your home desktop or mobile phone become quantum based anytime soon, just because quantum computers are at the early developmental stage doesn't mean they won't soon be affecting your life. The increased power and storage of quantum computers will be used by scientists to create even more accurate models and work on more complex problems, vastly improving research of all kinds. More immediately concerning, however, is the implication for data security. Modern computers use complicated mathematical codes to keep us safe. A modern supercomputer would take about a billion times longer than the existence of the universe to crack a standard 128-bit advanced encryption standard (AES) key, but a quantum computer may be able to do it in just a few moments. This is leading security experts to already reconsider how to keep us secure in the future.

SPEED READING

It's not just in storage where qubits shine. It may also be possible to use quantum entanglement to link two qubits together so that when one is a 0 the other is always a 1, or so that when the first is a 1 the other is also a 1, or really whatever combination you choose. This would mean that in reading a line of qubits you'd know what two lines say, cutting the time taken to read them in half and thereby making processing incredibly fast!

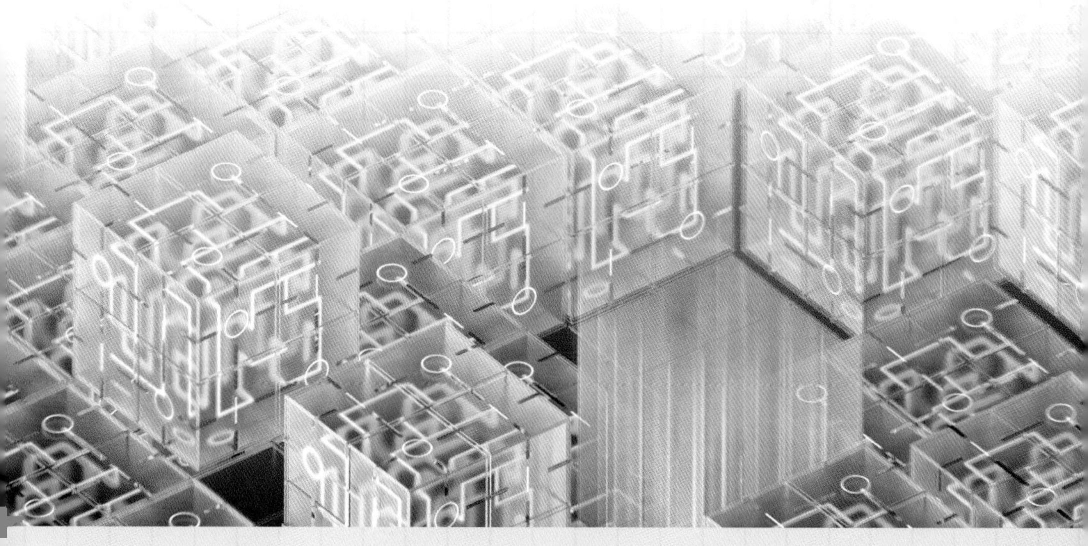

What is a pixel?

We look at pixels all the time on our computers, phones, and other devices, but do we really know what they are or how they work? A pixel (short for picture element) is the smallest element of a digital screen; these combine to create images.

Painting a Picture

Pixels are simple squares that can be made to be any color—but only one at a time. Much like in a mosaic, it is possible to then put many of these pixels together to begin to form an image. With just a few pixels it is only possible to make simple shapes, but as you increase the number of pixels the image begins to look more and more realistic. With a high enough number of pixels, it becomes impossible to distinguish them at all, only the completed image.

The Light Fantastic

The human brain is very easy to fool, especially when it comes to our vision. Pixels don't actually need to be able to produce the hundreds of thousands of colors we see, just three: red, green, and blue. Much in the same way that printers only need three inks and are able to mix them together into all the different colors you could want, pixels are able to use varying amounts of the three base colors to produce a dazzling array. Different types of screens work in slightly different ways, but one of the most common types is the liquid crystal display (LCD). A pixel on an LCD screen is made up of three subpixels: a red, green, and blue one. These are then covered by three colored filters, one for each color. Initially, the crystals in each of the filters are closely knitted together and block out most of the light from all of the subpixels. However, when an electrical charge is applied to a filter the crystals move apart, allowing more light to pass through. By carefully controlling the three different filters it is possible to generate literally millions of different colors, which can be used to create realistic images.

Resolution

Lots of people, especially advertisers, will tell you that when it comes to pixels, more is better. You may know some values for resolutions even if you don't know what they really mean. A common resolution for computer screens is 1,920 x 1,080; what this means is that from left to right there are 1,920 pixels and 1,080 from top to bottom, for a total of 2,073,600 pixels. The largest phone screens can be as pixel-heavy as 1,440 x 2,560, giving a whopping 3,686,400 pixels in all. It is also worth noting that a resolution only tells you how many pixels there are in a given space, not how big that space is. So, for example, a 1,080 x 1,920 phone screen will have a lot more pixels in the same space than a 1,920 x 1,080 computer screen, leading to a higher-quality image.

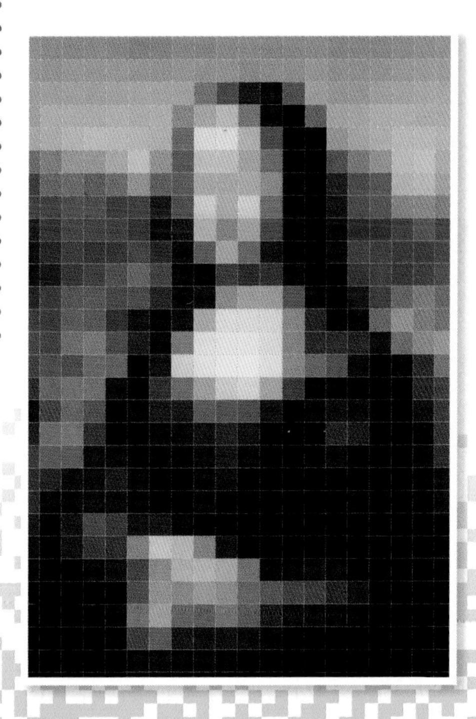

How do you recharge a battery?

Batteries are fairly expensive and disposing of them can be damaging to the environment. For these reasons, great efforts were made to discover a way that batteries could be used over and over. Rechargeable batteries use electricity to reverse the process that lets them create electricity on their own.

How Batteries Work

All batteries work in the same way. They consist of several parts, chiefly a metal section and another chemical called a reducing agent (such as zinc or cadmium). Batteries can be made of various combinations of metals and reducing agents. When connected to a circuit, the metal oxidizes, which changes the chemical composition, adding in oxygen, meaning (among other things) that it is able to release electrons. These electrons then flow around the circuit, producing the electrical current, and then flow back into the battery and into the reducing agent. Eventually, all of the metal oxidizes and there are no more electrons available to travel along the circuit and the battery dies.

Recharging

Rechargeable batteries are different, however, because by selecting specific materials for the metal and reducing agent (such as nickel-cadmium or lithium-ion) it is possible to make the reaction reversible. By applying an external electrical charge to rechargeable batteries it is possible to force the electrons out of the reducing agent back around the circuit and into the metal, causing it to become unoxidized and take in electrons. This means that the battery is ready to be used again. This is not a perfect process, however, and over time parts of the metal stop fully reversing, causing the battery to lose capacity as it has less metal to oxidize the next time.

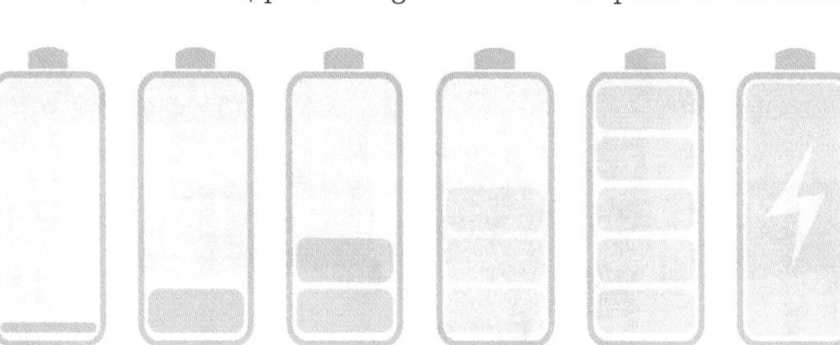

COMPUTERS AND ELECTRONICS

Think yourself a techie? A computer whiz? Then give it a test with this quick quiz.

Questions

1. What is the name of the future technology that will help hard drives become smaller?
2. What kind of waves does Wi-Fi use?
3. How many amps can car batteries produce?
4. What type of magnets are used in speakers?
5. What do fiber optic cables use to transfer information?
6. If there are fewer coils on the secondary wire in a transformer, what happens to the voltage?
7. Which kind of power generation is most unlike the others?
8. What is the name of the parts that store data in quantum computers?
9. Can you name two materials rechargeable batteries are made of?
10. How many colored elements do LCD pixels have?

Turn to page 214 for the answers.

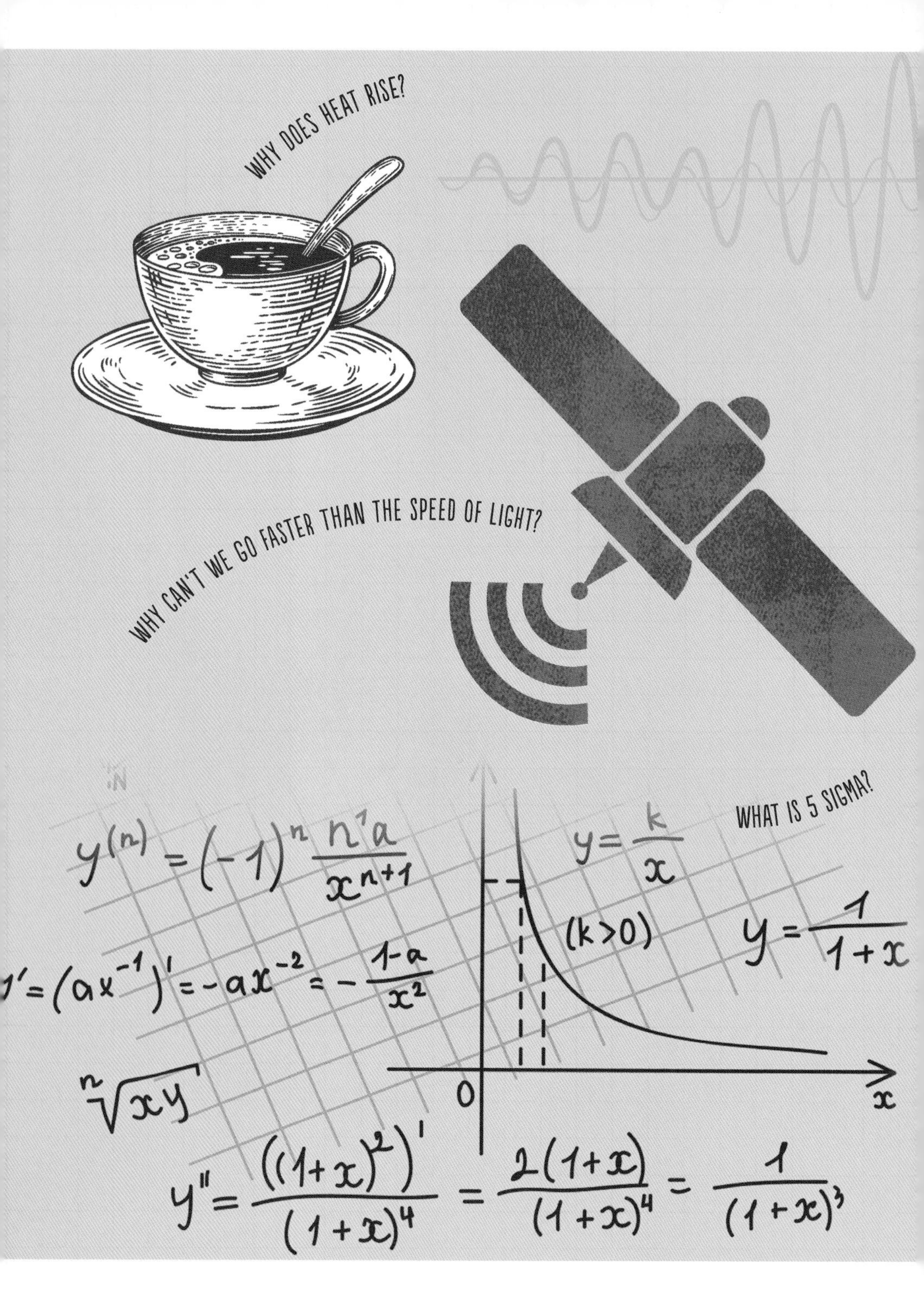
WHY DOES HEAT RISE?
WHY CAN'T WE GO FASTER THAN THE SPEED OF LIGHT?
WHAT IS 5 SIGMA?

FUNDAMENTAL PHYSICS

WHY DO CARS MAKE THAT ZOOMING NOISE?

HOW DO MAGNETS WORK?

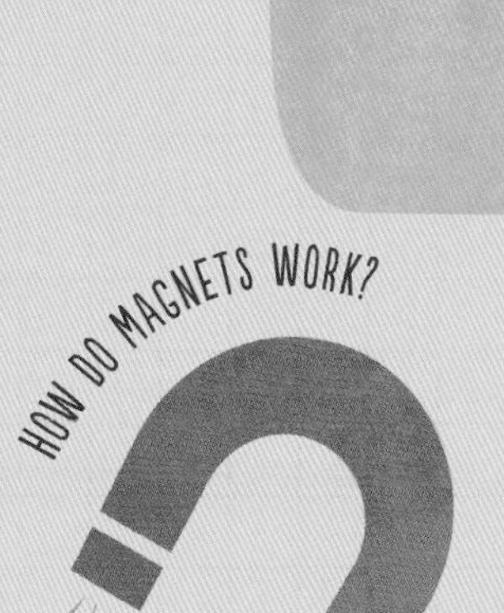

HOW CAN A CAT BE BOTH ALIVE AND DEAD?

Why can't we go faster than the speed of light?

There is a universal speed limit. Nothing can go faster than the speed of light (often written as "c"), which is 983,571,056 feet per second. It doesn't matter how many rockets you strap to something, you can't get any faster. This is because as something gets faster, it also gets heavier.

It's All Down to Inertia

This comes from Isaac Newton's principle of inertia, which, simply put, states that if you want to accelerate something, you need to use energy to do it. This is intuitive: If you give a wooden block a push across a table then you are using energy to make the block move. Then if you do the same but with increasingly heavy objects, they get harder and harder to move and you need to use more energy, until you get to something so heavy you can't push it. So the heavier something is, the more energy it takes to speed it up. We are limited by how much energy we can produce.

Increasing Weight with Speed

The full answer to why you get heavier when you get faster is complicated and involves some difficult math, but it can be simplified to the following equation:

$$m = m_0 \times \frac{1}{\sqrt{1 - \frac{v^2}{c^2}}}$$

What this equation tells us is that the mass of an object (m) is based on its "normal" mass (m_0), our speed (v), and the speed of light (c). At normal speeds that we experience in day-to-day life, v is so small compared to c that m and m_0 are basically the same. It's only when you get to about 25 percent the speed of light (still a whopping 245,892,762 feet per second) that the effects become even slightly noticeable. From this point, as the speed increases v^2 gets closer to c^2, which makes v^2/c^2 get closer to 1. This makes the number that m_0 is multiplied by bigger and bigger, until you reach a point where v=c, and this makes the number you multiply m_0 by infinity, meaning that m, your mass, will be infinite!

So if you want to accelerate an object past the speed of light, you will need an infinite amount of energy to speed up your infinitely heavy object. As this is not possible, it means we can't go faster than the speed of light.

SCIENCE FICTION EFFECTS

Traveling at super high speeds does some mind-bending things. Not only does it make you heavier, but if you were traveling very fast you would get shorter and your perception of time would slow down relative to the universe around you. This leads to all sorts of strange effects that you might think only matter in science fiction—but some of them are very real and need to be taken into consideration for things like GPS satellites and detection of deep space objects. Muons are short-lived particles that are made in our atmosphere by cosmic rays. They travel at near light speeds, but even going that fast they decay so quickly they should never reach the ground—the only reason they do is that because they travel so fast, the time they experience slows down, meaning they last long enough to reach ground-based detectors.

How can a cat be both alive and dead?

You've probably heard of Schrödinger's cat. It's a staple of science jokes and its use (or misuse) is commonplace. Less forthcoming, though, is an explanation of what it means. How can a cat be both alive and dead? Why does it matter? The explanation is tricky and quantum; it's all about how weird our universe really is.

The Infamous Cat

If you've somehow managed to avoid ever hearing of it before, Schrödinger's cat is a thought experiment first used in a 1935 paper by Erwin Schrödinger and it goes like this:

"A cat is locked up in a steel chamber, along with the following device (which must be secured against direct interference by the cat): in a Geiger counter, there is a tiny bit of radioactive substance, so small, that perhaps in the course of the hour one of the atoms decays, but also, with equal probability, perhaps none; if it happens, the counter tube discharges and through a relay releases a hammer that shatters a small flask of hydrocyanic acid. If one has left this entire system to itself for an hour, one would say that the cat still lives if meanwhile no atom has decayed. The psi-function of the entire system would express this by having in it the living and dead cat (pardon the expression) mixed or smeared out in equal parts."

This is to say that you put a cat in a box with a device that has a completely unknowable and random 50/50 chance of killing it. Then the quantum equation for this system will predict that the cat is both dead and alive (or perhaps a mixture of alive and dead) until the point at which you look into the box. Then it will be only one of those.

Alive AND Dead?

The absurdity of this idea was not lost on Schrödinger. In fact, the reason he posited this idea in the first place was that it was a ridiculous case that arises from the modern understanding of quantum mechanics being taken to its extremes. This all comes from the quantum idea of observation. In quantum mechanics, the very act of observing something will in some way change it. Specifically, this will cause a "waveform collapse." Any simple system can be written out as a mathematical equation known as a psi-function. Like any equation, it could have a number of different answers depending on what numbers you put into it. In quantum mechanics it's not until you observe the system that the numbers are actually put in. So before that point there is no true answer, only a mixture of all the possible answers.

WHY A CAT?

If you still haven't managed to wrap your head around it, don't worry. Quantum mechanics is not easy and often doesn't work in logical ways. Even the best scientists can still struggle. That's why Schrödinger's cat came about. It takes a very complex mathematical idea that can take years to study to a level of full understanding and breaks it down into a much simpler example. It might still be an unbelievably crazy idea, but then that's quantum mechanics for you.

What is the world's most famous equation?

If there's such a thing as a celebrity equation, then $E=mc^2$ is it. Originally penned in a slightly different form by Albert Einstein in 1905, it has become one of the most recognizable symbols of scientific achievement. It is important because it demonstrates mass-energy equivalence.

Matter Is Energy

Everything is made of atoms (see page 192), but if you look at it from a more fundamental viewpoint then you can break everything in the universe into two categories: matter or energy. All molecules, atoms, subatoms, and particles are matter. Matter is something that you might think of as physical and tangible, something that you could in some way "hold," and all matter has mass. All of the forces—electromagnetic waves like light, gravity, and the nuclear forces—are energy. Everything in the universe is either one or the other. The equation $E=mc^2$ shows that energy and matter are just different forms of the same thing, and that it's possible to turn one into the other. This is hugely important, because it means that everything in the universe is all different forms of one substance.

This can do some strange things: for example, if you took two wind-up alarm clocks that are identical, atom for atom, except that one of them had been wound up and was running, then it would have a greater mass than the other. This is because the energy stored in the spring, the energy in the moving hands, and even the energy of the heat caused within the clock all add a little bit of mass to the whole system.

Explaining the Equation

In Einstein's famous equation, E is energy, m is mass, and c^2 is the speed of light multiplied by itself. What this equation says is that it's possible to take an amount of mass (let's say 1 kilogram) and turn it into an amount of energy. The amount of energy that it can be turned into is equal to this mass multiplied by the value c^2. The value c is the speed of light in a vacuum and it is a known value at 983,571,056 feet per second. So c^2 comes out to roughly 300,000,000,000,000. This means each kilogram of mass can be converted into ten thousand billion joules of energy!

APPLYING THE EQUATION

In our daily lives, we don't see the effects of $E=mc^2$ because c^2, the speed of light, is so huge that the results are tiny, but mass-energy equivalence results in a lot of interesting effects. For one, it's the reason that we can't break the speed of light (see page 118). Understanding the relationship between mass and energy is what made it possible for scientists to invent nuclear fusion and weapons (see page 84). An individual atom will weigh less than the sum of its parts, as the mass is converted into a binding energy that holds it together. It is the breaking of these bonds that releases energy in a nuclear reaction. This missing mass also allows for the explanation of many different behaviors that atoms and even smaller particles exhibit.

> "Understanding the relationship between mass and energy is what made it possible for scientists to invent nuclear fusion."

Why do cars make that zooming noise?

The sound of cars traveling past is unmistakable; it's even more noticeable with race cars or emergency service vehicles with sirens blaring. It's not just cars, either—trains, planes, and basically anything moving past at speed makes the same zooming noise. So where does it come from? The noise is a result of the Doppler effect.

Doppler Shift

First, let's think of a stationary object giving off sound as waves that spread out in all directions around it. Assuming that the sound is constant, it would sound like a hum. When it moves, something strange will happen to the waves. As the object moves, it will continue to give off the sound waves, but the point at which they are initially made moves. This causes the gap between the waves to become shorter on one side (in the direction the object is moving) and wider on the other. This widening and shortening of the gaps in the waves means the frequency has changed. If an object is coming toward you, then the sound waves being closer together increases the frequency, shifting the pitch higher; if it's moving away, the longer distance between the sound waves shifts the frequency down, causing the pitch to get lower. The zoom that you hear from passing cars comes from this change in pitch of the car's sounds as it gets closer and then moves away.

Light Shift

It's not just sound waves that are affected by Doppler shift; it also happens with light waves. Objects need to be moving at a much higher speed in order for it to be noticeable, because the speed of light is that much higher than the speed of sound, but it still occurs. Objects moving toward you will have shorter wavelengths, making them appear slightly blue, and those moving away will have longer wavelengths, making them slightly more red.

EXPANSIVE IDEAS

It was by noticing that almost everything in our universe is red-shifted that Edwin Hubble came up with the idea that the universe was expanding. He also noted that the farther away things are, the more red-shifted they are (see page 45). These observations, which stemmed from the Doppler shifting of light, went on to be the founding principle behind the theory of the big bang.

This red and blue shifting of light is one of the best ways astronomers have of judging the speed and direction of distant objects in space. We know, for example, that the Andromeda galaxy is moving toward us at about 250,000 miles per hour because scientists are able to analyze how much bluer the light is than would be expected.

How heavy is a kilogram?

At first this might sound like a silly question. A kilogram weighs a kilogram, or perhaps you might say that it weighs 2.2 pounds. This doesn't answer the question of how anyone knows how much a kilogram should weigh in the first place. In short, a kilogram weighs a kilogram, but how heavy a kilogram is has changed.

> "A kilogram weighs a kilogram, but how heavy a kilogram is has changed."

Standardization

Before we look at the kilogram, we need to think about how the very idea of standardization works. If you want a kilogram of flour for a particularly large cake you are making, then you may either use a premeasured 1-kilogram bag or use scales to measure it out yourself. But the scales that you used would have had to first be calibrated using a kilogram. The kilogram used to calibrate must also be created at some point based on measurements of another, and so on. This means that everything must come down to one initial kilogram.

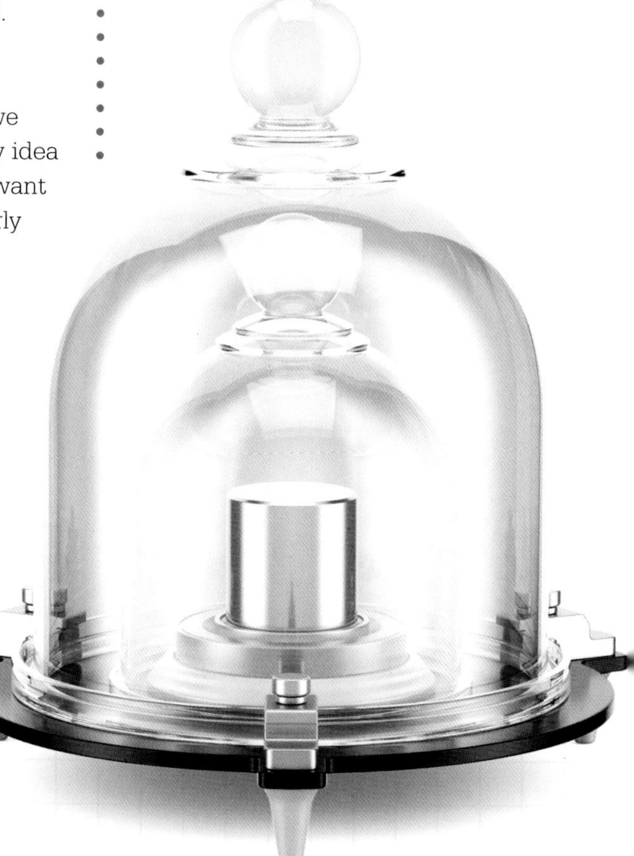

The International Prototype Kilogram

In 1889, how heavy a kilogram is was defined by the international prototype kilogram (IPK), a cylinder of platinum-iridium metal. Copies of it were made and then sent around the world to be the standard kilograms from which everything can be measured. There is a problem, however. Despite the fact that it's made of tough materials and kept in very stable conditions, over the last hundred or so years the kilogram has gotten heavier. By definition, the IPK is always 1 kilogram, but by comparing it against the copies, it seems that it has gotten heavier. So while it's still a kilogram, it just means that a kilogram today weighs slightly more than it did 100 years ago.

The Modern Kilogram

The weight of a measurement being able to change is, of course, not an ideal situation, but it is a flaw of any physical object. So in November 2018, scientists changed it so that the weight of a kilogram is fixed not to a metal rod but instead to a fundamental part of the universe. While the exact definition will be based on complex, theoretical equations, the physical form has a couple of different options. It could be based on a set number of atoms, say a sphere of silicon or carbon at a set size containing an exact number of atoms. The other way would be to set it against the amount of current required to lift it, so that weight is a measure of lifting current.

STANDARDIZING MEASUREMENTS

The kilogram is not by any means the first measurement to be defined using the fundamental laws of the universe. Many others are also defined in similar ways:

1 meter: The length of the path traveled by light in $\frac{1}{299,792,458}$ seconds.

1 second: 9,192,631,770 vibrations of a cesium atom.

1 kelvin: $\frac{1}{273.16}$ the triple point of water.

1 amp: The current at which two straight parallel conductors placed one meter apart in a vacuum would produce 2×10^{-7} newtons of force.

What is 5 sigma?

One of the central tenets of science is that you can never know anything with absolute certainty. There could always be something that you're missing or some brand-new idea that will trump it, or maybe just random chance will mean things aren't always as expected. So if you can't be 100 percent sure, then a new benchmark is needed—and that is 5 sigma.

Accuracy Probability

Much of modern science involves taking enormous sets of data and then analyzing them to see if anything interesting can be found. The problem is that there are many things that could mess with the data—temperature changes, tiny variations in the pressure, electrical interference, and even how long the equipment has been running could all introduce random statistical errors and noise into data. 5 sigma represents a probability of 99.99994 percent accuracy for a given phenomenon. That is to say that there is only a one in 1,666,666 chance that the observed effect is due to random statistics. Scientists used to use 3 sigma (a 99.73 percent probability), but several phenomena were later discovered to have been the result of noise in the data, so the more stringent 5 sigma is now used.

Standard Deviation

The term "5 sigma" means that something is five standard deviations above "normal." This comes from statistical modeling. Most experiments produce results that are randomly distributed around an average value. The standard deviation represents the spread of these random results. In a normal distribution, 68.2 percent of all of the randomness is contained under 1 sigma, 95.4 percent within 2 sigma. By the time you reach 5 sigma, this means that there is only a very tiny chance of that result having occurred through random chance.

How many dimensions are there?

The fourth dimension plays a prominent role in science fiction, and scientists often talk about additional dimensions to the ones we're used to. So just how many are there? Honestly, nobody is really sure just how many dimensions exist.

What Is a Dimension?

Before we consider how many dimensions there are, it's best to first look at what exactly a dimension is. It is best to think about a dimension as something that needs a coordinate. We can start with space dimensions, the ones we live in. We know there are at least three dimensions here because therc is up/down, left/right, and forward/back. If you want to meet up with somebody, you need to give them your position in the three dimensions. You also need to let them know what time you should meet at the agreed place, meaning you have an additional dimension you need to give. So our universe has four different dimensions: three spatial dimensions, and one for time.

Counting Dimensions

Mathematically speaking, there could be infinite dimensions, but the current suggestion is that our universe has four space dimensions, of which we only experience three (see page 45). Our universe could then exist as a four-dimensional part within a wider five-dimensional existence, and so on. Many theoretical physics theories, especially string theory, predict at least ten space dimensions, with some like the bosonic string theory going as high as 26 different space dimensions!

How do magnets work?

If you've ever played with a couple of magnets, you might have been confounded and fascinated by the invisible force between them that can feel like magic. They work, quite simply, by creating magnetic fields that are able to interact with each other.

Permanent and Electro

There are two different types of magnets that generate their magnetic fields in different ways. Electrons are charged particles that move around in an atom, thereby generating a magnetic field. This means everything creates a magnetic field. However, in most objects the direction of the field made by each individual atom is randomly distributed, so they all effectively cancel each other out. In a permanent magnet, however, all of the tiny fields point in the same direction, causing a big magnetic field. The other type of magnet is an electromagnet. Passing electricity through a wire causes electrons to move through it and thus a magnetic field is generated. By coiling up the wire, it is possible to increase the magnetic field to a usable strength. Crucially, electromagnets can be turned off and on at will.

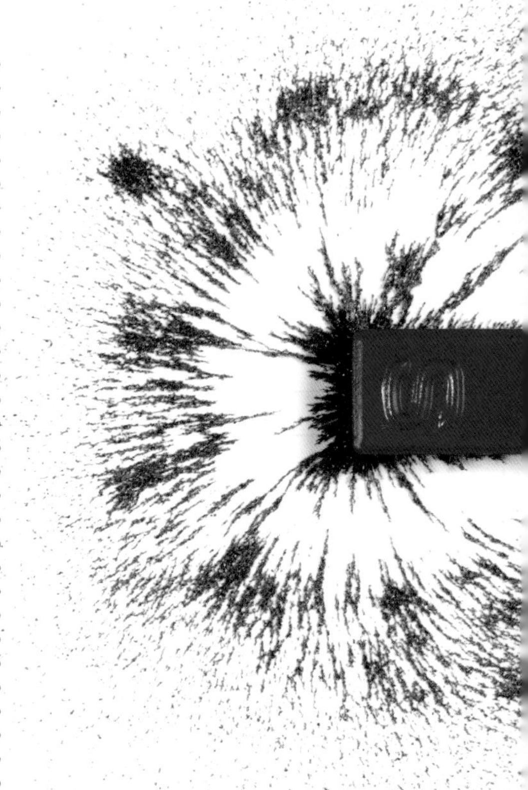

Magnetic Fields

Fields are a strange quirk of physics. They probably don't really exist and are just a very good way of visualizing an abstract concept. Fields are made up of field lines; these are like flowing rivers of force that can push or pull in one direction. A magnet will give out these field lines, which get weaker and weaker the farther away they go from

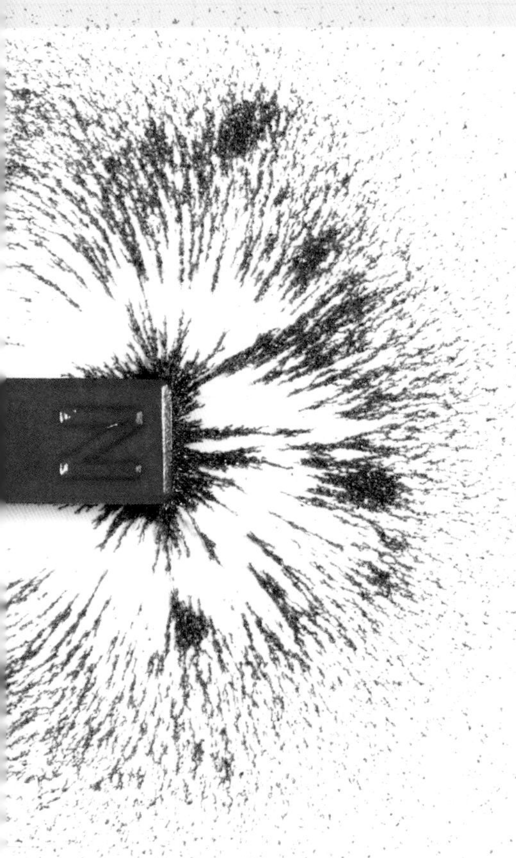

the magnet. Magnetic fields always flow from a positive (north) charge to a negative (south) one. The magnetic force we experience is a result of this. If you have two magnets, try putting their positive ends together. The force from the field lines will flow outward from both magnets, pushing them apart. Alternatively, moving opposite ends will align the field lines, causing the magnetic force to pull inward. Magnetic fields have another interesting property in that magnetic lines can never cross. Attempting to make them do so results in a lot of repulsive force, which adds to the already strong effect of the field lines.

MAGNETIC MONOPOLES

All magnets have a "north" side and a "south" side. If you cut a magnet in two, all you end up with is two magnets. Why isn't it possible to make a magnet that has just a north pole or a south pole? Why can't we make a magnetic monopole? Nobody is really sure; in fact, there is no physical reason that magnetic monopoles shouldn't exist—and modern theories do allow for them, but despite many attempts and some initial success, as of yet none have been found or created.

Does time go faster at the top of the Eiffel Tower?

Relativity is a complicated topic—it took the genius of Albert Einstein to figure it all out. Its discovery completely revolutionized physics and everything we thought we knew. One of the major things that it left in its wake was our idea of time as unchangeable. So, does relativity mean that time really is different at the top of the Eiffel Tower?

What Is Relativity?

If I were to ask you what speed you're traveling in a car, you might think it's easy to answer, but it's not. The speedometer might read 40 mph but that's only a measure of how quickly you're moving relative to the ground. If there was a car next to you going 30 mph, then relative to that you're only traveling 10 mph, or to a train going past at 90 mph, you're actually going at -50 mph.

Relativity is the idea that no one of these answers is more correct than any other—it's all relative. You might think that it's the movement compared to the Earth that is "right," but don't forget that the Earth itself is orbiting the Sun at over 60,000 mph, and why would the Earth be more "correct" than the Sun?

What Has This Got to Do with Time?

Another aspect of relativity is that the speed of light is always the same. This is regardless of the relative motion of whatever you're looking at. To wrap

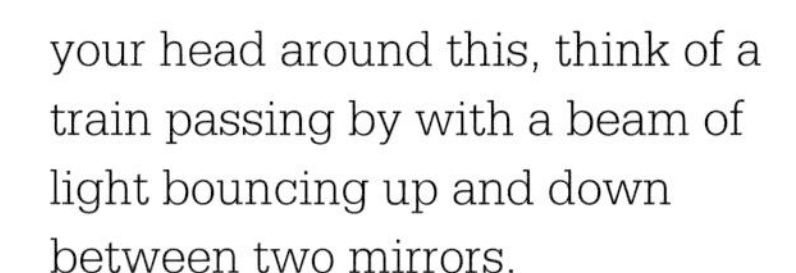

your head around this, think of a train passing by with a beam of light bouncing up and down between two mirrors.

If you were on the train, you'd see the light bounce up and down at the speed of light (c), and it takes a certain amount of time. If, however, you were standing by the tracks and watched the train pass, you would also observe the light traveling at c.

Now if you're standing by the track you see not only the light traveling up and down but also to the side as it moves with the train. This means you see the light travel over a greater distance, but because the speed of light has to be the same, then there can only be one conclusion. You observe time itself on the train moving slower.

Gravity and Relativity

The final piece of the puzzle is that because space and time are all wrapped up together in space-time, the effects of high speeds are also present in gravity. That is to say that gravitational fields also cause space to slow down.

This means that, living on Earth, we are subject to a constant gravitational force that slows down our time relative to anybody outside. The higher up you go, the weaker the gravity gets and the faster time gets, meaning that at the top of something like the Eiffel Tower, time does in fact go faster. Normally these effects are so small so as not to be noticeable, but things like GPS satellites have to take time slowing into account or they would quickly become useless.

Why can't we untoast bread?

This is perhaps one of the most important and fundamental questions we can possibly ask. In answering it, we explore the very nature of the universe itself: where it came from, where it's going, and how exactly it will get there. This question (it may not surprise you) isn't really about toast. But we can't untoast bread because of entropy.

Once Toast, Always Toast

When you toast bread, a number of things occur. There is a series of reactions as the amino acids and sugars are reacted together, causing them to mix and brown. The sugars in the bread can also caramelize, which makes them melt and spread out. All of this increases the entropy in the toast and, because entropy only ever goes forward, you can't undo it.

The Arrow of Time

Entropy is defined as "the measure of a system's thermal energy per unit temperature that is unavailable for doing useful work." To the lay reader, however, this might not be particularly useful in understanding what it actually is. It is often called a measure of disorder and it has one key property: Entropy is always increasing. Think of a block of ice dropped into a pot of hot water. At the start, the ice atoms are relatively ordered—cold in one part, hot in another. But over time the temperatures will become the same and the atoms originally in the ice will melt and mix with the rest in the pot. Over time, the disorder in the system has increased.

Entropy is sometimes called (somewhat grandiosely) the "Arrow of Time," because it is the only thing that can only ever go in one direction, always getting bigger. At the beginning of the universe, everything was in a near-perfect order. Since then it's been getting more chaotic, the amount of disorder has been ever increasing, and every process that happens in the universe serves to increase the amount of entropy.

Order vs. Complexity

If entropy is only increasing, and things are getting less ordered and more chaotic, then how can organized things like stars, planets, and even human beings form? Just because entropy overall must always increase, that doesn't mean it isn't able to decrease in some places. It's like how making ice in a freezer will decrease the entropy in the water but the freezer will proportionally increase entropy outside the device. Complexity, however, is very different from order. In our previous example, as the ice and water begin to mix, it becomes increasingly complex, as to define the system properly would be increasingly difficult, until eventually it all becomes just water again—so complexity arises as part of the overall increase in entropy. In a sense, our existence is probably an effect of entropy.

> "Because entropy only ever goes forward, you can't undo it."

Why does heat rise?

We often talk of heat flowing from one place to another, and we also say that it rises without considering what heat really is—and by extension how it does it. Heat moves as it does because it works as a transfer of kinetic energy.

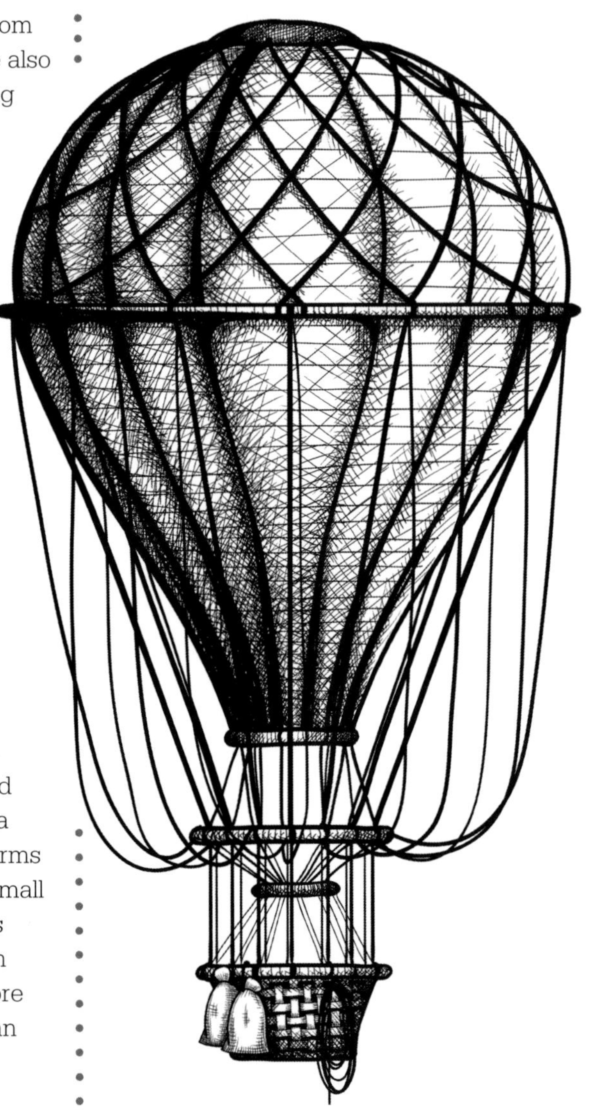

What Is Heat?

We often talk about heat as if it were some kind of fluid, but temperature is not an entity in its own right; rather, it is a property that matter has. If we take air as an example, it's full of trillions of molecules of gas. Each molecule is moving around, and the speed that it moves at is its temperature. The more energy the molecules have, the faster they move, and thus the temperature is higher. This should not be confused with the wind, which happens on a much larger scale; movement in terms of temperature is only using very small distances at a molecular level. This increased movement also occurs in liquids and solids, but it comes more from vibrations and interaction than from free movement like in gases.

How Heat Moves

If an individual molecule gains more energy, it will move faster and over a larger distance; this means that it has a greater chance of bumping into and interacting with another molecule. In doing so it will transfer some of its energy to the other molecule, which could then itself pass this on. This interaction happens over and over again, causing the bouncing molecules to spread away from the other highly energetic molecules. This is why heat always flows from hot to cold; it is the transfer of energy from the source outward.

On the Up

Heat rises because warmer molecules move around more, and this makes them more spread out than the colder ones. This means that they are less dense than the cooler ones. This causes them to float upward as the colder, denser molecules sink through. This results in the shimmer you can see on hot roads or over radiators—the heated surface gives the air molecules around it additional energy that makes them hotter. Being hotter, the air becomes thinner and less dense, which allows it to rise. Because it's thinner, it holds less water than the surrounding air, causing distortions in the passing light—this is what gives hot air its distinctive shimmer.

EQUILIBRIUM

If a warm molecule bumps into a cold one, then it will give it some energy; if it bumps into a hotter one, then it might take some energy. Eventually the system of molecules will reach a point where all of them have exactly the same amount of energy; this is equilibrium. It's the reason hot drinks cool down to room temperature or ice cream heats up to room temperature; they all reach equilibrium.

What is the correct way to make a cup of tea?

How to make the perfect cup of tea is a subject of much debate. Such questions are best answered, of course, by science. We must assume that our tea is made using a tea bag with just a splash of milk and no sugar.

It's All About Heat

In blind taste tests, it has been determined that hotter tea tastes better. So the real question is: Does adding the milk before or after brewing affect the temperature of the resulting tea?

When you first make some tea, it will be at about 212°F. Then, as you let it brew, it will start to cool off. Adding milk will also induce rapid cooling as the colder liquid mixes in. Both of these cooling effects are going to happen regardless of when the milk goes in. Temperatures tend toward equilibrium (see page 135); that is to say something hot will give its heat to its surroundings until it reaches the same temperature. How quickly this happens is dependent on the temperature difference. The larger the difference, the faster the heat is transferred.

When to Add Milk

In a cup of tea, by adding the milk before letting it brew, you cause an initial cooling of the now milky tea, which will cause the overall liquid to cool less than the tea would alone in the same amount of time. This means that when you add the milk to the tea after the brewing, it will become colder and thus taste worse. (The actual difference is only a couple of degrees and is not really noticeable, but it's still scientific.)

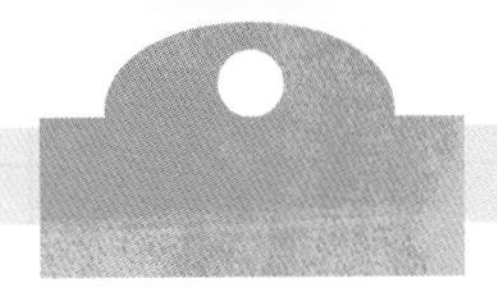

FUNDAMENTAL PHYSICS

Got the basics down? Test yourself on these questions about the most fundamental parts of physics.

Questions

1. What happens as you get faster?
2. What kind of equation can be used to describe any system?
3. What does the E in $E=mc^2$ mean?
4. What type of shift happens to an object moving toward you?
5. What level of sigma used to be used?
6. What material is the international prototype kilogram made of?
7. How many time dimensions are there?
8. What are the two kinds of magnets?
9. What is the principle that means time can go faster and slower?
10. What is the scientific name for the "Arrow of Time"?

Turn to page 215 for the answers.

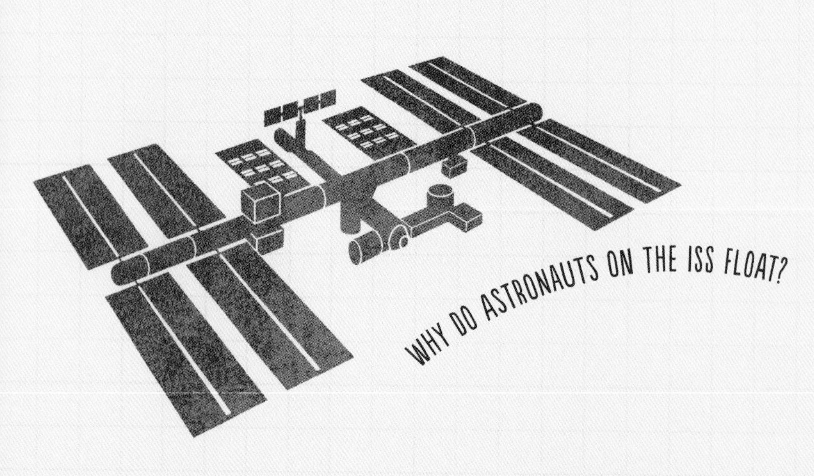
WHY DO ASTRONAUTS ON THE ISS FLOAT?

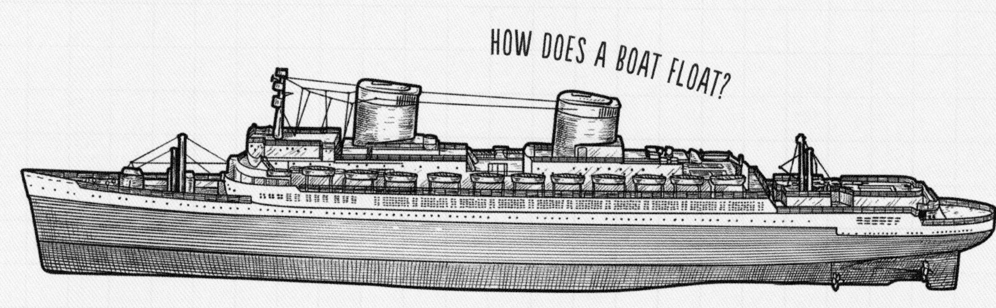
HOW DOES A BOAT FLOAT?

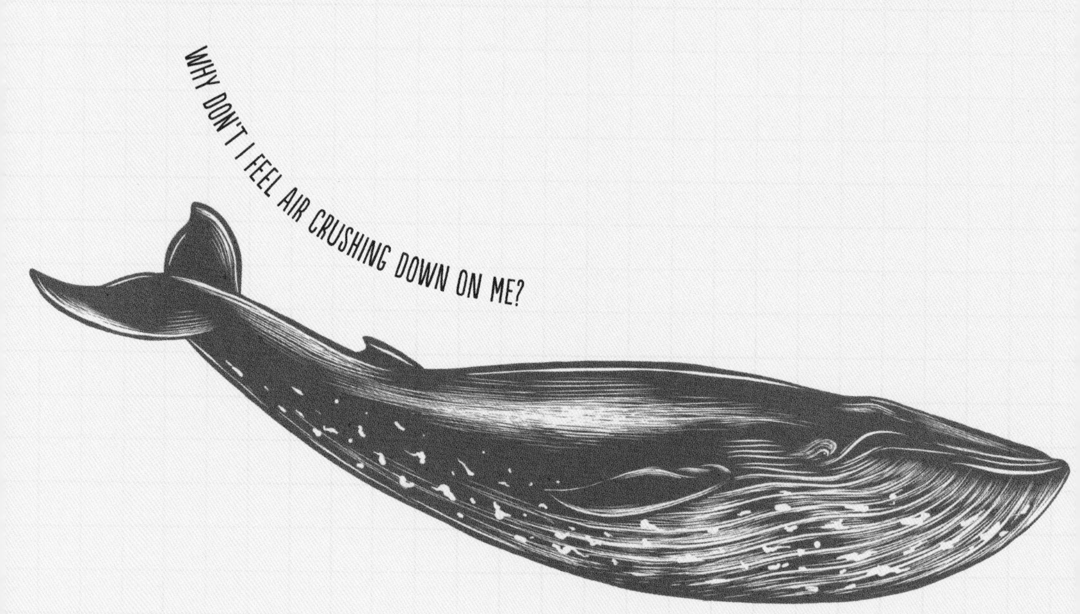
WHY DON'T I FEEL AIR CRUSHING DOWN ON ME?

HOW LONG WOULD IT TAKE TO FALL THROUGH THE EARTH?

FORCES

HOW FAST IS GRAVITY?

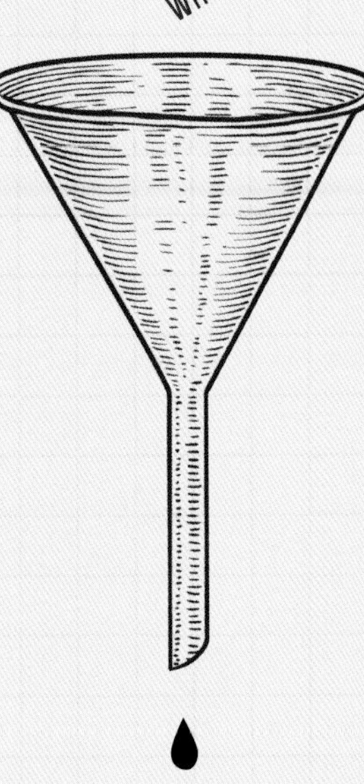
WHY IS KETCHUP HARD TO POUR?

Why do astronauts on the ISS float?

The International Space Station (ISS) is a satellite orbiting 254 miles above the Earth and is the only inhabited space off the planet. If you've ever seen footage of the astronauts inside it, you can't have helped but notice that they're floating around. The reason they float is because they're falling.

Gravity Is Everywhere

Gravity pulls us toward the ground. Because the astronauts inside the ISS out in space are floating around, many people believe that they aren't being affected by gravity, but this is simply not the case. Gravity as a force never stops, and it is produced by everything in the universe. There is gravity

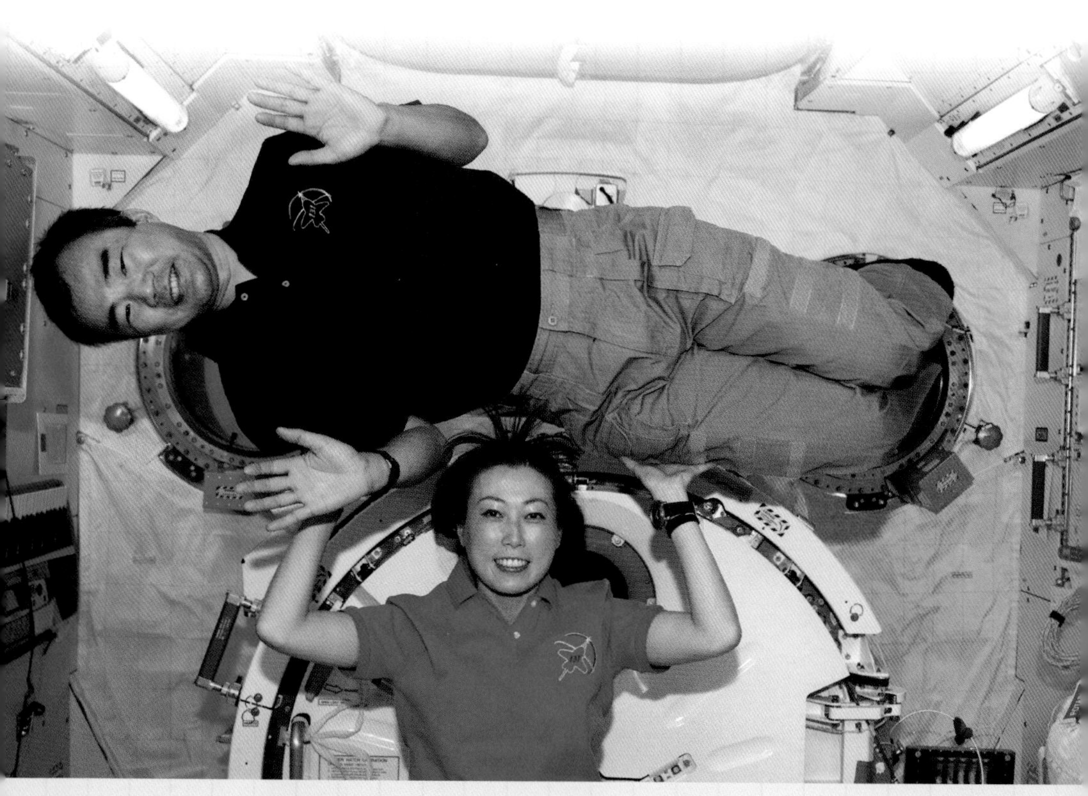

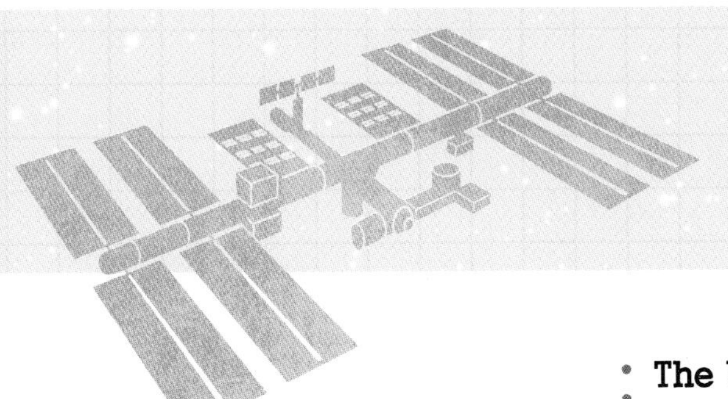

everywhere—even the most distant and emptiest parts of the universe still experience gravity. Gravity does, however, get weaker the farther it travels, so it can be effectively zero. This is still not the case for the ISS, though. In fact, gravity on the ISS is still 89 percent the strength of gravity on Earth, so it's not about gravity.

Always Falling, Never Landing

Okay, so maybe it is about gravity a bit. Earth's gravity is still pulling on the ISS, and it is pulling it down toward the ground; however, the satellite has enough sideways momentum that instead of being pulled into a crash landing, it's always missing and is instead being pulled around the Earth in an orbit. So essentially, the ISS is constantly falling, which means that everything and everyone in it is, too. Because they're all falling at the same speed, this effectively means that it is as if no gravity is acting on them, because they are in constant freefall.

The Universal Forces

There are four fundamental forces in the universe:

The Strong Nuclear Force: The force that holds atoms together.

The Weak Nuclear Force: The force involved in some nuclear interactions.

The Electromagnetic Force: The force that is made by charged particles.

Gravity: The force made by mass.

Gravity is by far the weakest of all of the forces (which is why you're not constantly stuck lying on the ground), but it's one of the most important because it interacts on such a massive scale and is produced by everything in the universe! The more massive an object is, the bigger its gravitational pull, which is why the Sun can hold planets in orbit with gravity, but you can't hold anything at all with it. However, gravity (like all the other fundamental forces) is also dependent on distance. The closer two objects are together, the greater the gravitational attraction between them. One of the cool things this means is that when you hover your hand just above somebody else, you're exerting just about as much gravitational force on them as the Sun is!

How does a boat float?

Heavy things sink, lighter things float; it makes sense intuitively. And yet boats are big and heavy, but they manage to float just fine. So there's clearly more going on here than at first glance. Boats are able to float thanks to the help of the buoyancy force.

Floating Boats

In order to make boats float, this buoyancy force is exploited. If an object is fully submerged, then the amount of liquid it displaces is equal to the volume of the object. However, as the boat is being placed in the water and becoming submerged, the buoyancy from the displaced water will cause the boat's weight to decrease. It decreases based on its submerged volume and the density of the liquid. Boats have large hulls that go underwater; they are designed specifically so that their volume displaces enough water so that the buoyancy force is equal to the gravitational force pulling down. This means that the boat as a whole has no net force up or down, making it able to float.

Buoyancy

Gravity pulls down, and objects that are more dense sink below the lighter material. This is why if you put a rock into water, it will sink. However, there is also an opposing force that comes from the water itself called buoyancy, which comes from water displacement. You may have heard of this phenomenon through the (likely apocryphal) tale of Archimedes. He supposedly got into a full bath, causing the resulting displaced water to slosh over the side. In the story, he was then able to use this to calculate the density of a crown. This is because when you submerge an object, it displaces an amount of liquid based on its volume (which can be used with its weight to calculate density), and this displacement causes a force that pushes back upward; this force is called buoyancy.

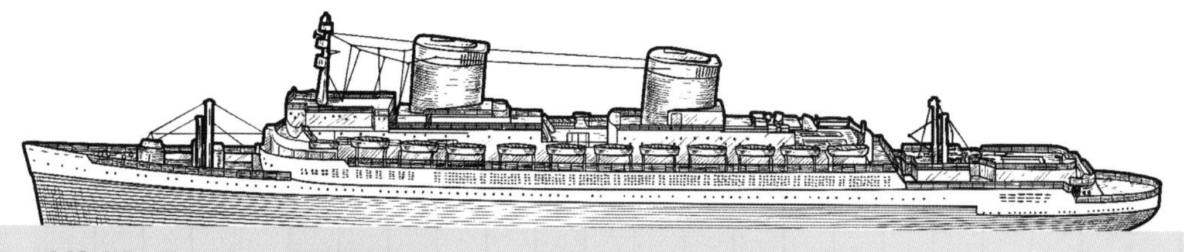

Why is it possible to hammer a nail into wood but not push it in?

Imagine you're building a shelf or other bit of carpentry and you need to get some nails in. There's only so much force that you can provide with your hands, so why does using a hammer work where pushing the nail won't do more than mark the wood? It all comes down to impulse.

Impulse

When you apply force to an object, the result of the applied force is affected by the duration of time that the force is applied. Impulse is the measure of the average force across the time that it's applied. If force is applied to an object over a shorter period of time, then the impulse is lower and there is a greater effect. It's why sometimes shoving an object will move it where pushing it wouldn't—because the force is applied over a shorter period. So by hitting the nail with a hammer rather than trying to push it, there is a greater resultant force on the nail.

Torque

Hammers also use another aspect of physics to help increase force. You might have noticed that it's easier to hammer a nail when holding the hammer farther down the handle. This is because of torque. Torque is the resultant force generated by a rotating object, and it multiplies the used force by the distance between where the force is applied and the point of rotation. In swinging a hammer, you pivot the hammer in your hand and the head is some distance away from your hand, meaning the force hitting the nail is greater. The farther down the handle you hold it, the greater the distance between the pivot and the head, so the greater the resulting force.

Could a coin dropped from a tall building really kill somebody?

Gravity causes objects to be attracted to each other and causes them to accelerate. The old story goes that if you dropped something hard like a penny from a high enough height, it would build up enough speed to kill somebody standing below. A penny would certainly be able to get fast, but because of terminal velocity, a dropped coin couldn't kill anyone.

Acceleration

Gravity is an acceleration force that pulls toward the ground: this means that falling objects are caused to accelerate. Gravity on Earth has a strength of about 32 ft/s^2, which means that every second the speed of a falling object will increase by 32 feet per second. So after one second, it will be traveling at 32 feet per second, after two it will be at 64, and so on. Using the Empire State Building at a height of 1,200 feet, it would take a penny dropped from the top just under 9 seconds to fall the whole way, and by the time it reached the ground it would be traveling with a speed of about 280 feet per second, which is about 190 miles per hour!

Air Resistance

Gravity is not the only force acting on the penny, however. As it falls it will bump into lots of air molecules, which causes air resistance, slowing the acceleration down. This force acts to balance out gravity. As the penny gets faster, the air resistance increases until the two are balanced and the coin has reached its maximum speed, known as

terminal velocity. As pennies are very light they flip and turn as they fall, which causes them to encounter even more air resistance. This means they have a low terminal velocity at only around 75 miles per hour, so if the penny landed on your head it would hurt but wouldn't be deadly.

The heavier an object is, the greater its momentum, which means that even at the same speed its impact will be harder and more dangerous. Heavier objects are also more likely to drop straight down, reducing any additional air resistance, meaning they can reach higher speeds than things like pennies. This is why objects such as nuts and bolts falling from scaffolding can be dangerous—hence the reason construction workers wear hard hats to protect themselves.

FELINE FALLING

It is a well-known fact that cats can survive high falls, and they do this by exploiting terminal velocity. As a cat falls from a height, it twists and turns onto its front and spreads its body out. Doing this increases its surface area, which in turn increases its air resistance, reducing the terminal velocity to a nonfatal level. There is a small catch to this, however, in that the cat needs to be several stories up in order to have the time to flip properly and flatten itself out, meaning that more cats die falling from the first few floors than those above—there are reports of cats surviving falls of up to twenty-six stories!

How can a leaf stop a train?

If you've ever had a train canceled on you, then you know it's very frustrating. It can be even more frustrating to be told that it's been canceled because of leaves on the line. While this might seem like a silly reason, leaves can actually be very dangerous because of friction.

Friction

Friction is a resistive force that is generated when any two surfaces come into contact. When you try to slide one object over another and it stops, it's because of the friction. Low friction means that the two things can move more easily. Friction is very important to motion; the friction between your feet and the ground is what allows you to walk on it with ease. A lack of friction can be difficult to deal with, as you might know if you've ever slipped on ice or snow.

Leaves on the Line

Train tracks are manufactured to a fine balance: There is a low level of friction between the tracks and the wheels, so that the train can move easily along them (if the friction were high, it would require a lot more fuel to move the train), but not too low, as the train must be able to stop safely. When leaves fall and get damp, they can be sucked onto the tracks by a passing train and then crushed into a watery paste by the wheels. This results in the leaves turning into a slippery, oily layer that covers the tracks. This is a very dangerous situation, as it makes it harder for trains to slow down and stop. In order to account for this, the trains need to move slower and more carefully, leading to delays and cancellations.

How fast is gravity?

Gravity is a force that pulls on us all the time from everywhere. So if it's affecting us from everywhere, does that mean gravity takes time to travel, like light or matter? It was originally thought to be instantaneous, but it was eventually discovered that gravity travels at the speed of light.

Travel Time

Everything, even fundamental forces, takes time to travel. If the Sun somehow instantaneously vanished, then it would take about eight minutes for us to see it happen here on Earth. This is as true for light as it is for electromagnetism as it is for gravity (see page 116). This is important on a fundamental level because it also limits the speed at which information can travel. Nothing can affect anything else quicker than the speed of light.

Hypernova

One of the strangest effects of the fact that it takes gravity time to travel is a hypernova. This occurs when a truly enormous star dies. The center of the star undergoes a collapse and turns into a black hole. But because of the time it takes for the effects of the collapse to travel to the outside of the star, it will still continue to shine as though nothing had happened for a short while. The outside of the star then catches up and is dragged into the black hole, creating an incredibly bright event some ten times brighter than a normal supernova.

Gravity travels as gravitational waves, as first predicted by Einstein. When an object with mass moves, its motion produces a wave that distorts space-time like a wave in the sea. However, as gravity is very weak it takes massive events like colliding neutron stars to actually see them.

Why don't I feel air crushing down on me?

You may not have ever stopped to think about it, but there's a lot of stuff above your head. There are miles and miles of air, dust, and water. Literally tons of stuff is resting above your head—so why don't you feel it? It's because of the balancing of pressures.

A Rhinoceros or Five

Stand in an open space and there is air being pulled down onto you by gravity. The atmosphere above us weighs about 1 ton per square foot, which is like five rhinoceroses stacked on top of each other. Obviously, if you had five rhinos balanced on your head you'd feel it, so why do you not even notice the air? Air is a fluid—well, it's a gas, but it acts as a fluid. It's able to flow and change shape. This means that all this weight is distributed evenly in all directions as opposed to just straight down. So while there might be the weight of one rhino pushing down on you, there is also another rhino pushing upward, meaning that they cancel each other out and you hardly feel anything at all.

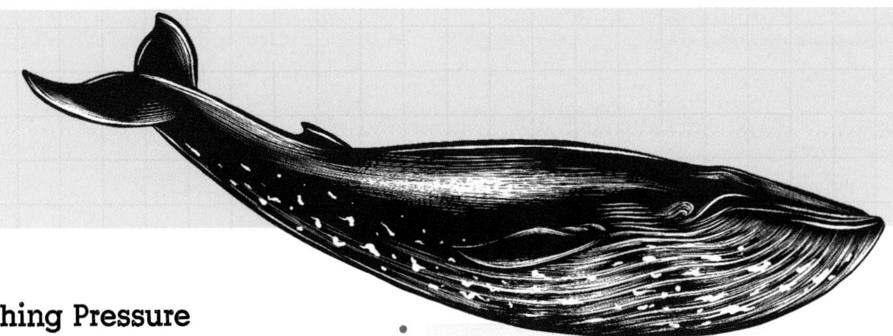

Crushing Pressure

It's easy to demonstrate what happens when you mess with this delicate balance. If you take a plastic bottle and suck the air out of it, it will start to collapse inward. This isn't because you're somehow pulling it inward with the sucking but rather because you're removing the air inside so there is no longer a balance of pressure, meaning the external air is able to crush the bottle.

What About Water?

We are able to go about our daily lives without worrying about the air pressure around us, but if we try to dive deep underwater, it becomes important to think about. As you dive deeper, the water above you stacks up and subjects you to greater pressures—like with the air, but the difference is that this pressure can kill. If water is a fluid, what makes it different? The answer is that us humans simply aren't adapted for it.

The average air pressure on your body is about 100 kPa (kilopascals). Our bodies are naturally able to withstand this pressure, but can struggle in much higher pressures. As you dive underwater, the pressure begins to increase. Roughly every 33 feet farther down, the pressure increases another 100 kPa. Diving organizations suggest going no deeper than 131 feet, as it is below this depth that the water pressure reaches dangerous levels for humans at above 500 kPa.

UNDER PRESSURE

Many animals are better adapted for high pressure. Blue whales are able to dive to depths greater than 330 feet. In 2014, scientists captured a sea ghost sailfish on camera at 26,715 feet below sea level, meaning that it is able to withstand pressure on its body of nearly 82,000 kPa! Submersibles have been made to allow humans to dive deep into the ocean with several reaching the bottom of the Challenger Deep, the deepest known point in the Earth's seabed hydrosphere, at just under 36,000 feet.

Why is it quicker to make a cup of coffee at the top of Mount Everest?

Going up Mount Everest to make a cup of coffee is not recommended. You would need to contend with the arduous climb, the snow, and the bitter winds. But if you were able to make it, you would notice that your water would start to boil a lot faster than normal. This is because of the lower air pressure up there.

Boiling Points

We all know that water boils at 212°F, but that's actually only the temperature at sea level, or rather under normal atmospheric conditions. As you get higher, the air pressure drops and with it so too does the boiling point of water. On the top of Mount Everest's 29,029-foot peak the air pressure drops to just 34 kPa (from a standard 100 kPa), meaning water could boil at just 160°F.

Not a Great Cuppa

Boiling is the process by which a liquid is turned into a gas. This happens when the vaporization pressure in a liquid (which increases along with temperature) is equal to the atmospheric pressure around it. This is why water can boil at a lower temperature on the mountain. This does mean, however, that the cup of coffee you make will only be at 160°F, and it would therefore likely take longer to brew because of the lower temperature. So if you do want a good cup of coffee, it's probably more time-efficient to stay down the mountain.

How does sound travel?

Sound has to get from point A (the source) to B (where it is heard). Much like light, it does so by moving as a wave. Unlike light, however, it isn't able to move on its own. Sound travels by causing a moving vibration through some medium.

Vibrations

All sound starts as a vibration. In human speech, it comes from the voice box; in speakers, from a moving disk (see page 104), but whatever the source, it is started as a vibration. This vibration is then transferred to the medium it's in contact with. Usually this is air; the vibrations make the air molecules next to the sound source jostle and move, mimicking the vibrations, which makes the air next to those molecules do the same, and so on, allowing the sound to travel through the air. Sound travels through liquids in much the same way, making the molecules vibrate to transfer it along. In solids, the molecules don't move around, but they are still able to vibrate. In fact, because the molecules are neatly ordered, solid materials carry sound best, though human ears are more adapted to hear in the air.

In Space, No One Can Hear You Scream

Space is a vacuum, which means it contains nothing (or close enough to nothing). Because space is empty, it means there is nothing for the sound to travel through. This is probably a very good thing, because if it could, then the Sun would constantly be making a deafening noise at about 125 decibels, which is louder than the sound of thunder directly overhead. It does mean, however, that all of the epic space battles you see in movies with their lasers and rockets and huge explosions would, in reality, be completely silent.

How long would it take to fall through the Earth?

Let's say it was possible to build a hole directly through the center of the Earth and out the other end. Then let's say that for some reason you decided the best idea would be to jump into it. If you undertook such a crazy endeavor, then you would indeed be able to pop out the other end, and it would take 42 minutes and 12 seconds.

Journey Past the Center of the Earth

Calculating something like this can be a little tricky, so to simplify we would have to assume a couple of things—namely that the Earth is a perfect sphere of equal density and that our hole has no air in it. When you jumped into the hole, you would start to fall and every second you would speed up by 32 feet per second (see page 144). You'd soon be traveling very quickly and getting deeper and deeper into the hole. However, as you got deeper, the pull of gravity would start to get weaker. As you got farther into the Earth, there would be less gravity below you pulling down and more above pulling up—you'd still be getting faster, but you would be doing so more slowly. Eventually you would reach the center of the Earth, where there would

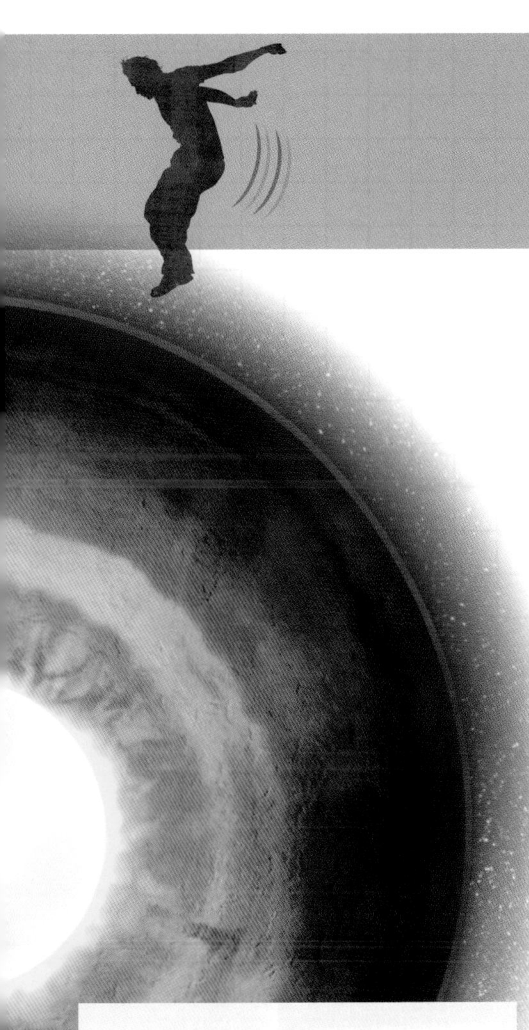

be basically no gravity affecting you at all as the Earth's gravity would be pulling you equally in all directions. However, as you would now be traveling very, very fast, you would shoot right through the center and carry on. Because there would now be more Earth above you than below you, there would be a greater pull upward, which would start to slow you down until eventually, by the time you popped out of the hole on the other side, your speed would have reached zero.

Not Just the Center

Using some math to work out the journey tells us that because of the gravitational pull of the Earth, this trip would take 42 minutes and 12 seconds. However, crucially, it is the gravitational pull that tells us this, not the distance. If you dug a hole between any two points on the Earth—not just two directly opposite each other—that would form a straight line through the middle, so long as you were able to fall through it unimpeded then the journey would always take the same amount of time. Of coursc, if you did fall right through the center of the Earth, you would be burned to a crisp.

DIGGING HOLES

It is worth pointing out that the idea of digging such a hole is not that practical. The Earth is on average 7,918 miles thick and the deepest that has ever been drilled is the 9-inch-diameter Kola Superdeep Borehole, which only reaches a meager 7.6 miles down into the Earth's crust.

Why is ketchup hard to pour?

If you've ever had to get ketchup out of a glass bottle, then you know that it can be a real problem. You might shake and slap, hold and wait, or just opt to go without, as doing too much risks having it all come out at once, ruining your food. The reason ketchup can be such a difficult substance is because it's a non-Newtonian fluid.

Non-Newtonian Fluids

All fluids have a property called "viscosity." It's kind of like the friction of the molecules in the fluid as they rub against each other. Liquids with a low viscosity, like water, are able to flow easily, but liquids with higher viscosity, like syrup, pour much more slowly and are far more sticky.

Non-Newtonian fluids don't just have one viscosity, though. Their viscosity is based on the force being applied to the liquid. A blob of ketchup has a very high viscosity—it's more like a lump of slime than a liquid. It's only when you apply some force to it that its viscosity lowers, letting it flow easily.

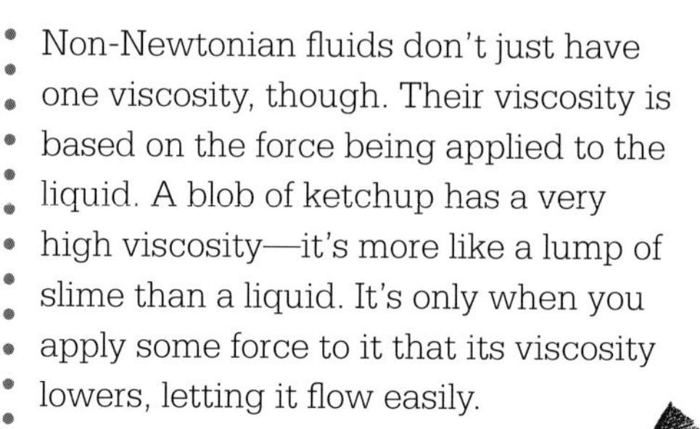

THE MOST VISCOUS

The most viscous fluid is pitch (a material often used for surfacing roads). In 1927, Professor Thomas Parnell at the University of Queensland in Australia set up an experiment where he poured hot pitch into a funnel, let it settle for a few years, and then recorded the drops as they formed. It took just over eight years for the first drop to fall, and in 2014 the ninth drop fell. It is expected that the next drop will fall at some point in 2028. This makes pitch nearly 280 billion times more viscous than water.

FORCES

Are you a force to be reckoned with when it comes to answering questions? Show that this quiz is nothing but a pushover.

Questions

1. What is the name of the force that would cause your bath to overflow if you filled it to the brim and then got in?
2. What percentage of Earth's gravity do astronauts on the International Space Station experience?
3. If you used a hammer to push a nail into some wood, what effect would you be taking advantage of?
4. What is the name of a falling object's maximum speed?
5. What force prevents your feet from slipping on the ground?
6. What forms at the center of a hypernova?
7. How many rhinos' worth of air is crushing down on you all the time?
8. At what temperature does water boil on Mount Everest?
9. What does sound travel through the best?
10. What type of fluid is ketchup?

Turn to page 215 for the answers.

WHOSE DOG (POSSIBLY) HELD BACK SCIENCE?
WHICH PHYSICIST WAS PROSECUTED FOR HERESY?
WHOSE NOBEL PRIZE NEEDED TO BE MADE TWICE?
WHO WAS THE FIRST PHYSICIST?

WHO WAS THE OWNER OF THE MOST DANGEROUS NOTEBOOK?

PHYSICISTS

Aut. Sculp.

DID EINSTEIN REALLY FAIL MATH?

Did Einstein really fail math?

It's a statement that's often repeated to struggling math students—that even one of the world's smartest men failed the subject at school and was a bad student. But is this really the case? It may not surprise you to learn that the man who figured out some of the most difficult problems of his age was not only a decent student but did not fail math.

Young Einstein

As a very young child, Albert Einstein was perfectly normal; it doesn't seem he started talking late (something else attributed to him) and he probably didn't have a learning disability. In his earliest school days he did well but wasn't thought of particularly highly and he seemed to struggle with the learning process and even with teachers themselves. This aside, by the time he was eleven years old, he was already reading university-level texts

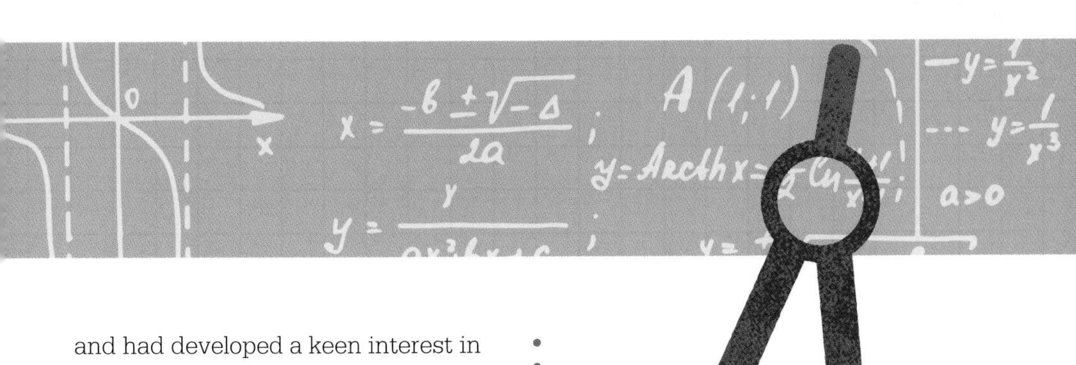

and had developed a keen interest in physics. So where does the idea that he failed math come from? Einstein did fail an entrance exam to the Zurich Polytechnic, which he took two years early, though this was because of his poor results in French and natural studies rather than his math ability.

How Smart Was He?

Everyone knows that Einstein was a genius, but it's hard to appreciate how incredible he was. In 1905, Einstein was just 26 years old, had just received his PhD, and was working in a patent office. In this year he published four groundbreaking papers on completely different topics. Any one of them would have made him one of the most important scientists of his generation. Over the course of his life, Einstein managed to make huge contributions to many of the major fields of physics, publishing over 300 papers, and remains to this day one of the most respected scientists of all time.

Einstein's Achievements

Einstein's greatest contribution to science was probably his theory of relativity. This is the idea that there is no single reference frame for our universe and that everything is relative. It took him ten years to get it right, but the publication of his work in 1915, which set out the mathematics for it, stands as the biggest fundamental shift in how we view the universe since Isaac Newton's paper *Philosophiæ Naturalis Principia Mathematica* in 1687.

Einstein also managed to explain the occurrence of the photoelectric effect, which led to the acceptance of the wave-particle duality of matter (see page 196); described Brownian motion, which led to the adoption of atom theory; and discovered mass-energy equivalence, as defined in his famous equation $E=mc^2$ (see page 120), which in turn led to the development of nuclear energy, including atomic bombs.

Whose dog (possibly) held back science?

If you've ever had a dog, you'll know that they're lovable, loyal, and more than a bit clumsy. Sir Isaac Newton's dog set fire to his early manuscripts on gravity.

Diamond Damage

Newton was a dog lover and had a Pomeranian called Diamond. One evening, while working by candlelight on his burgeoning theory of gravitation, he stepped out of his office for a time. While he was gone, the dog managed to bang into the table covered in his papers, which knocked over the candles and started a small fire. Although the fire didn't spread and damage much of the rest of the room, it did manage to burn all of Newton's papers. Supposedly, upon his return to his office Newton said to his dog: "O Diamond, Diamond, thou little knowest the damage thou hast done." Newton then fell into a depression from which it took him months to recover, and it was a year before he was able to reproduce his lost ideas. It should be noted that this story may not be true, much like the story of Newton's apple, but in a way it's reassuring to think that even some of the greatest people in history can be subjected to such misfortunes.

The Masterpiece Prevails

Despite this setback, Newton was able to publish his masterpiece *Philosophiæ Naturalis Principia Mathematica* in 1687. In this work he set out the idea of universal gravitation, showing how the planets and their moons move. It was the final nail in the coffin for the idea that the Earth was the center of the universe, and formed the basis of our understanding of the universe today.

OUTSTANDING CONTRIBUTIONS

It wasn't just gravity that Newton worked on. He dabbled in alchemy, the nature of light, and a host of other topics. Newton's greatest contributions to science, however, were not strictly speaking discoveries in themselves. Newton created calculus, a mathematical notion that remains today one of the best ways for us to explore the natural world. Its importance cannot be understated; much of the physics that underpins our modern world is a result—one way or another—of calculations that use calculus. Perhaps even more importantly, Newton helped to define the purpose of scientific inquiry. He was the first to state that the ultimate goal of physics was to find the underlying laws of the universe and understand them. It was also his input that finalized the formation of the modern scientific method, as described in his statement that for a theory to be correct, it must completely match up with the observable universe; any discrepancy that could not be explained must render it wrong.

Who was the first physicist?

This isn't as easy a question as you might think. Technically Isaac Newton was the first, as before this point physics was just part of the wider study of "natural philosophy." That, however, is an unsatisfying answer, as physics had been studied long before then.

Thales of Miletus

The first person who could be considered a physicist was Thales of Miletus (c. 624–546 BCE), who was born in modern-day Turkey. He was a philosopher, astronomer, mathematician, and sage who studied widely. He was, of course, not the first person to question the world he lived in or why it was so—this was a great Greek tradition—but the unique way he went about it is what makes him stand out as the first physicist.

Thales proclaimed that every event must have some natural occurrence. While this may seem obvious to us now, the ancient Greeks intrinsically linked their worldview to their gods and put much thought into the influence of the gods in worldly affairs.

Thales expanded on this idea by attempting to categorize the night sky, finding the thing that everything was made of and crucially giving predictions of what he might find.

The Hellenistic Scientific Revolution

While the idea of physics may not have been solidified until the era of Newton

(1643–1727), the ancient Greeks made lots of discoveries that have formed the basis of all scientific ideas that we find today.

ARISTOTLE (c. 384–322 BCE) created the concept of logical thinking to determine if something can be true. He made some of the first attempts to explain motion and the composition of matter. His work was hugely influential and became a leading school of thought well into the time of Galileo (1564–1642), and still remains important to this day.

ARCHIMEDES (c. 287–217 BCE) was a great inventor who began to apply mathematics to the physical world. This opened up a whole new way to explore physics, and let him predict things like buoyancy and forces with incredible accuracy.

PTOLEMY (c. 100–170 CE), pictured right, took the ideas of observation and theory from Aristotle and combined them with Archimedes's mathematics to create a model of the universe. While this model, with the Earth at its center, was very wrong, it allowed for all of these elements to be married together.

IF I HAVE SEEN FURTHER...

It is very difficult to pinpoint who can really be called the first true physicist. Even Issac Newton did not use all the tools that we would today say make a physicist. Physics is a process whereby the work of many people across the ages builds up to let us know more about the world we live in. As Newton wrote to a friend: "If I have seen further, it is by standing on the shoulders of giants."

Which physicist was prosecuted for heresy?

Major religions have not always been open to change. In centuries past, the Catholic Church had very real power beyond the spiritual guidance of its people—and they defended it rigorously. So in 1633, when Galileo Galilei was seen to be openly mocking the wisdom of the Church, he was brought to trial for heresy.

The Center of the Universe

Since near the start of the Common Era, Ptolemy's theory that the Earth was the center of the universe and that everything else orbited around it was widely accepted. This didn't sit right with many people, including Ptolemy's peers before and after him, but with the decline of Western science as the Roman Empire fell and the golden age of Islam focusing its efforts elsewhere, a rise in fundamentalist Christianity and its interpretations of the Bible meant geocentricism took hold.

By the early 1600s, however, the evidence against geocentricism was starting to mount up. Nicolaus Copernicus's mathematically based model for the solar system, published in 1543, placed the Sun at its center; this was backed up in 1609 by Johannes Kepler's detailed investigation into the motion of the planets, leading to the rise of heliocentrism (literally meaning "sun in the middle"). When Galileo used the newly invented telescope to look to the sky and observed the moons of Jupiter orbiting around the planet and not the Earth, he had made up his mind.

Shaking Things Up

In 1610 Galileo published *Sodereus Nuncius*, which outlined his theory that not everything orbited the Earth. It proved controversial, but its findings

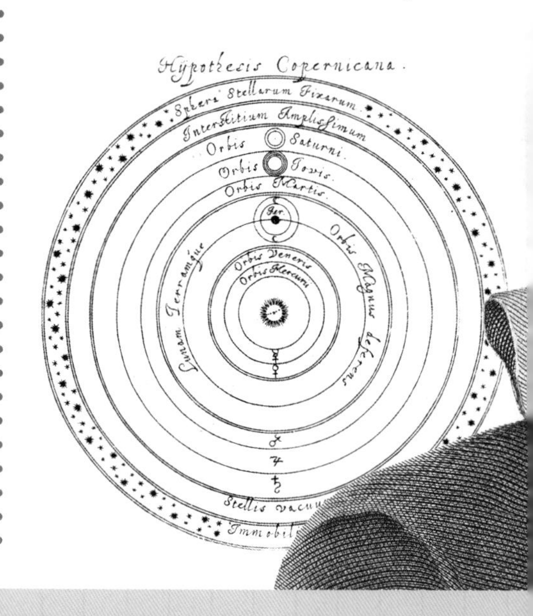

A LEGACY OF HERESY

Many scientific ideas have been denounced as heresy by different religions (though few as formally as heliocentrism). Continental drift, germ theory, blood circulation, and evolution have all received significant pushback from religious groups throughout history as science has clashed with Scripture. This is not to say that religion is opposed to science, as even in Galileo's time many cardinals studied his works and defended his ideas. The political leanings of organized religions have, however, sometimes caused them to take opposing positions against science.

were solid and could be repeated easily by others. It was here that he first angered the Catholic Church. The text was brought before the Vatican's Inquisition on charges of heresy. Despite Galileo's attempts to defend it, the book was formally declared heretical, as it contradicted Scripture. The book (along with many similar works by others) was banned and Galileo was told to abandon his heliocentric beliefs.

Several years later, Galileo decided to kick the hornet's nest by releasing a second book, *Dialogue*, in 1632. The arguments contained within it were heavily critical of the theory of the Earth-centered universe, and it painted the defendants of such a model in a very poor light, describing them as simpletons. So in 1633, Galileo was summoned once again before the Vatican's Inquisition, this time to personally stand trial for heresy. The evidence was damning and the Church was under significant pressure to punish heretics, so the trial concluded with a guilty sentence. The punishment was life imprisonment, which was swiftly commuted to house arrest.

Who was the owner of the most dangerous notebook?

When we think of harmful scientific equipment, we might think of giant lasers, enormous magnets, or deadly chemicals—but probably not a simple notebook. And yet there is a scientist's notebook that, nearly 100 years after it was last used, is so dangerous that it needs to be kept in a lead-lined box. That notebook belonged to Marie Curie.

Unseen Danger

Marie Curie was the first person to do detailed research into the radioactive properties of certain elements. While she didn't know it at the time, nuclear radioactivity is incredibly dangerous and comes in several forms (see page 206). Her laboratory was filled with radioactive samples of polonium and radium, and she even carried bottles of the stuff in her pockets much of the time. One of the dangerous properties of radiation is that if it's strong enough and exposure is long enough, it's possible for it to cause other objects to become irradiated and to start giving off radiation themselves. Almost all of Curie's

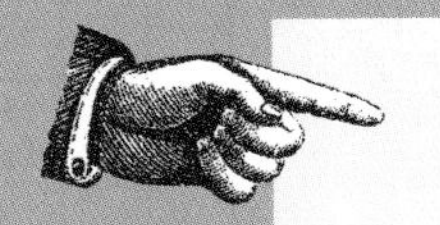

WORLD WAR I AND BEYOND

Curie's contributions went beyond theoretical study. During World War I she became the head of the Red Cross's radiology division and created portable X-ray machines to help diagnose injured soldiers. It is estimated that over a million soldiers benefited from her equipment. After the war, she focused much of her time on raising money and awareness of radioactive research, giving lectures across the world and playing an important part in many high-profile scientific bodies before her death in 1934.

possessions became completely irradiated by the substances she used. Curie eventually died of radiation-related illness. Her body was so radioactive that her coffin had to be lined with an inch of lead to prevent it from leaking out and potentially harming others.

It's a Man's World

It is no secret that the history of physics is dominated by males, with only the occasional woman managing to make her mark. It was in this masculine-skewed world that Curie had to prove herself. As a woman she was unable to gain access to higher education and so she joined the underground "Flying University" that secretively taught women in Poland. She eventually managed to earn a place at the University of Paris, where she lived a life of poverty, studied hard, and earned two degrees. It was there that she met her future husband and work partner, Pierre. She returned briefly to Poland but was denied a place at Krakow University on account of her sex and settled again in Paris.

Curie resolved to explore the newly discovered X-rays. Even the lack of a proper laboratory didn't put her off—she used a converted shed next to the school at which she taught. Curie took extreme care to ensure that it was clear that the research was her own and not that of her husband, as she realized how hard it might be for some to accept that such work could be done by a woman. Eventually she was recognized for her work and received a Nobel Prize in Physics with her husband and another scientist for their work on radiation, and later received a Nobel Prize in Chemistry.

Whose Nobel Prize was given to somebody else?

These days science is a collaborative effort. Gone is the era of lone, eccentric physicists in their home laboratories writing up grand treatises. Now groups of people, tens or even hundreds, work toward solving a problem together. The decision of who gets the credit for discoveries therefore becomes a little more tricky—and because of this, Jocelyn Bell Burnell missed out on a Nobel Prize.

Her Discovery

In 1967 Jocelyn Bell Burnell was working as a PhD student under Antony Hewish. She spent two years helping to build a 4.5-acre radio telescope made of poles and wires. Once it was completed, it produced more than 100 feet of paper charts per day, which Burnell had to analyze.

At some point she noticed a strange mark in the data. After several months of checking for equipment error, and taking multiple and more detailed readings of the same point in the sky, it was still there. From this she concluded that what she was seeing in the data was definitely real and in space. She used her telescope to find several of these strange signals in different parts of space, and they all produced the signal in a regular pattern. The signals were later classified as evidence for pulsars and received enough interest and recognition that the 1968 paper that announced them to the world, "Observation of a Rapidly Pulsating Radio Source," was chosen for Nobel Prize consideration. In 1974 Antony Hewish and Martin Ryle (a fellow academic) received the Nobel Prize in Physics "for their pioneering research in radio astrophysics." Burnell herself, who first discovered the pulsars, did not earn recognition.

Who Gets the Prize?

Nobel Prizes can be awarded to up to three people, so if hundreds helped on a discovery, then who gets it? The Nobel Prize often goes to the head scientist and largest contributors on a project, though it can be more than a little political. Many feel Jocelyn Bell Burnell was cheated of a Nobel Prize, as she was the first to discover pulsars and pushed for further investigation. However, Burnell herself stated:

> "I believe it would demean Nobel Prizes if they were awarded to research students, except in very exceptional cases, and I do not believe this is one of them . . . I am not myself upset about it."

LITTLE GREEN MEN

Pulsars are very small and dense stars that are highly magnetic. This causes them to fire out enormous beams of electromagnetic radiation. They spin at very high speeds, which causes their beams to move around, much like a lighthouse's. The effect of this here on Earth is that telescopes observe the beam as a flashing signal from the sky. The signal from pulsars is more regular than all but the best of atomic clocks. While Burnell and the team were attempting to determine the source of the mystery signals, the idea was briefly considered that such a perfect signal may be from some alien civilization. While this idea was quickly dismissed, the nickname "Little Green Men" (LGM) stuck around until the explanation of their origin replaced it with the name pulsars.

Which physicist died from politeness?

Nobody likes to be rude. But some take it much further than others, going to great lengths to avoid annoying others. Tycho Brahe, however, went a step too far and ultimately died because he did not want to breach etiquette.

A Polite, Unpleasant Death

On October 13, 1601, Tycho Brahe attended a banquet. During the meal he found himself in need of the toilet, but as to leave the table would have been rude, he stayed. On returning home he found himself no longer able to urinate—this then likely caused his bladder to burst, killing him.

The Work of Brahe

Brahe was gifted an island and titles by the Danish king, and on this island he built a castle dedicated to the arts and sciences in which he manufactured a great many scientific instruments, and it became a hub of learning and experimentation. Brahe was an exceptional astronomer: His equipment allowed for observations leagues beyond anything else of his day. He created catalogs of the stars and huge sets of data that were referred to for years to come in the blooming scientific revolution. His work became a standard for empiricism and precise, objective scientific measurement. Beyond this, he also discovered supernovas and was able to prove that the stars are much farther away from Earth than was previously thought.

TYCHO'S NOSE

By all accounts, Tycho Brahe was a character. Tales of his antics could fill volumes of their own, and his revelry was renowned. In his university days he quarreled with another student (supposedly over some mathematical formula), which resulted in a sword duel between the two of them. He was scarred on the forehead and lost the end of his nose; he wore a fake brass one for the remainder of his life.

Whose failed experiment was a huge success?

In physics, as in life, not everything works out. But sometimes it's the failures that can lead to more important changes than they would have had they been successes. So it was that the failure of Albert A. Michelson and Edward W. Morley's ether-finding experiment set the basis for a revolution in physics.

Luminiferous Ether

The "luminiferous ether" was the name of a supposed substance that scientists believed filled all of space. It was thought to be the medium through which light travels, much in the same way as we now know that sound travels through air (see page 151). The idea of this ether was important, as it was believed to provide a single frame of reference for everything in our universe.

Michelson and Morley (pictured right) set out to measure the luminiferous ether. To do so they set up a giant cross with a half-silvered mirror in its center and a mirror on two adjacent arms. A single light beam could then be sent down one arm into the half-silvered mirror, which would split the beam in two, sending the beams into separate mirrors, which would then bounce them back into the silvered mirror and down the final arm, where a detector was. Here, they believed, the effects of the ether would cause the two light beams to vary slightly, leading to a fringe pattern on the detector.

They Found . . . Nothing

However, no result was found; the experiment was a dud. Despite tightening up the equipment, changing and upgrading it, trying it at different times of the year and multiple repeats, there was just nothing to be found. More experiments by others some years later looking for the luminiferous ether also found nothing. While this disheartened them, the lack of any proof for the ether meant that when Einstein's theory of relativity came to the fore, it was quickly adopted and ushered in a new era of physics.

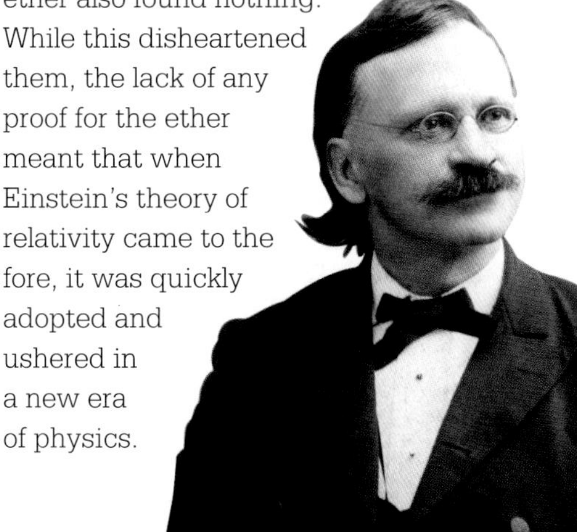

Which physicist broke into U.S. nuclear facilities?

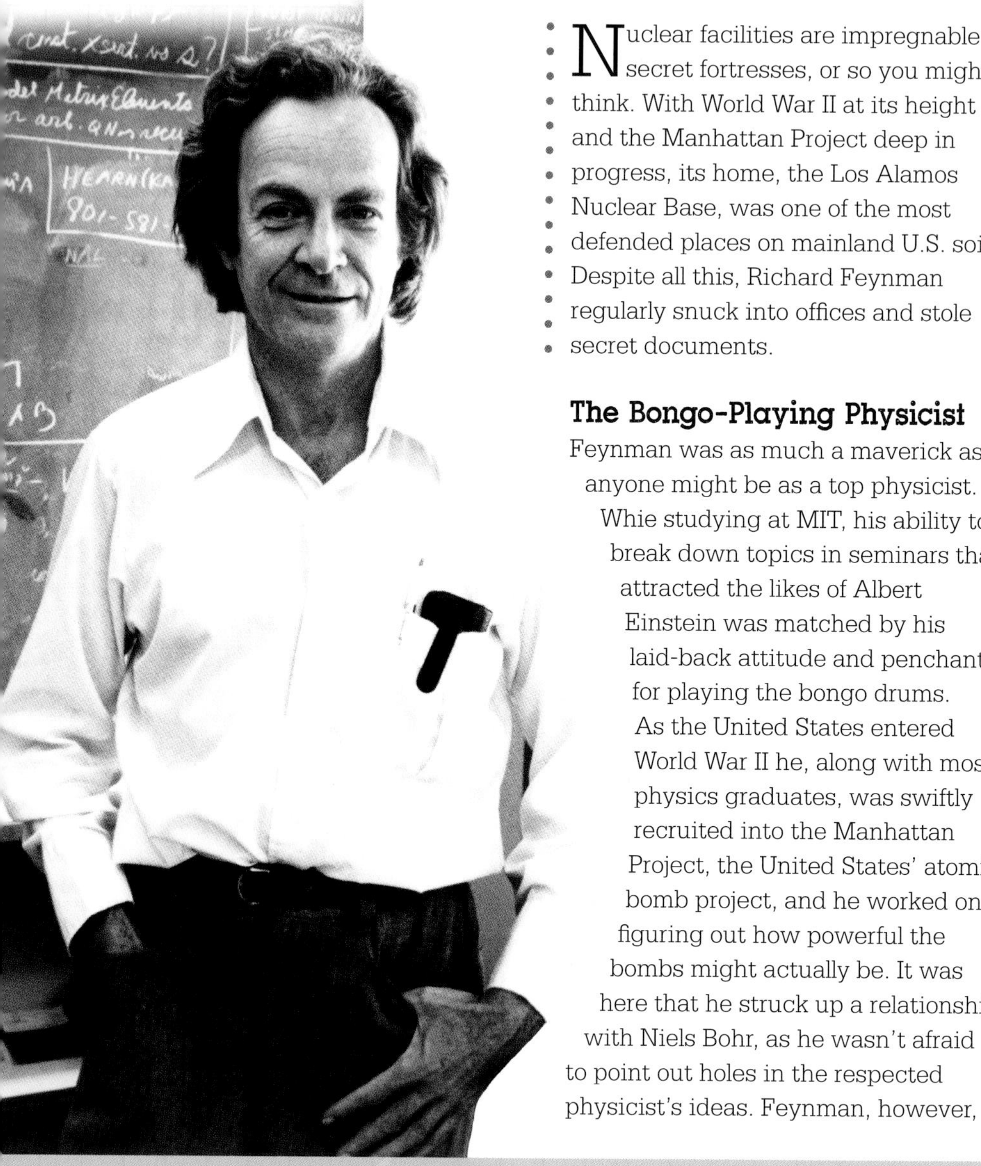

Nuclear facilities are impregnable secret fortresses, or so you might think. With World War II at its height and the Manhattan Project deep in progress, its home, the Los Alamos Nuclear Base, was one of the most defended places on mainland U.S. soil. Despite all this, Richard Feynman regularly snuck into offices and stole secret documents.

The Bongo-Playing Physicist

Feynman was as much a maverick as anyone might be as a top physicist. Whie studying at MIT, his ability to break down topics in seminars that attracted the likes of Albert Einstein was matched by his laid-back attitude and penchant for playing the bongo drums. As the United States entered World War II he, along with most physics graduates, was swiftly recruited into the Manhattan Project, the United States' atomic bomb project, and he worked on figuring out how powerful the bombs might actually be. It was here that he struck up a relationship with Niels Bohr, as he wasn't afraid to point out holes in the respected physicist's ideas. Feynman, however, soon grew bored working in the remote facility and so began to entertain himself by breaking into his colleagues' filing cabinets to take reports he needed for his work when they weren't around. Security was soon informed and installed better locks, which he also managed to crack. Feynman himself said:

> "I opened the safes which contained all the secrets to the atomic bomb: the schedules for the production of the plutonium, the purification procedures, how much material is needed, how the bomb works, how the neutrons are generated, what the design is, the dimensions of the entire information that was known at Los Alamos."

THE FEYNMAN LECTURES

Feynman took a teaching post at the California Institute of Technology (Caltech) in 1952. He became a relatively popular speaker and was asked to rework his lectures. Between 1961 and 1963, he gave the most famous series of lessons ever known: the Feynman Lectures. Delivered with his trademark charisma and approachable style, the lectures were audio-recorded and written up as a collection of books. These remain among the best-selling physics books of all time and are essential reading for any aspiring physicist.

Feynman's Diagrams

Feynman's work was largely based around the interactions of subatomic particles. It was a tricky area that involved a lot of mathematics and complex ideas. Unhappy with how the problems were set out, Feynman developed the Feynman diagram. The diagram allows for a simple representation of a particle interaction, showing the incoming particles, how they interact, and what comes out of the other end, including whether there are any antiparticles (see page 202).

Whose Nobel Prize needed to be made twice?

A Nobel Prize is the most prestigious award a scientist can receive. It shows that the recipient has led great research into the unknowns of the world. Some incredible people have had the honor of receiving more than one Nobel Prize. Rarer, however, is the occasion when a Nobel Prize needs to be made twice, such as in the case of Niels Bohr.

A Scientist on the Run

Niels Bohr was awarded a Nobel Prize in 1922. He lived in Denmark, and through the 1930s he and many others helped refugees fleeing persecution during the rise of the Nazi regime in Germany. He offered financial support, refuge, and temporary jobs to a number of high-profile scientists escaping Germany and found them places to move to in other parts of the world. After Germany occupied Denmark in 1940, word reached Bohr that he was to be arrested, so he fled to Sweden.

Before he did so, however, he instructed his friend George de Hevesy to dissolve his Nobel Prize, along with those of several scientists who had fled Germany, in a mixture of nitric and hydrochloric acid to hide them from the Nazis. The bottle of liquid sat on a shelf in the Theoretical Research Institute of Copenhagen until after the war, when it was retrieved and the metal was extracted before being recast into Nobel Prizes once more.

A PRIZE MODEL

Niels Bohr was awarded his Nobel Prize for his planetary model of the atom showing electrons orbiting a central nucleus, which is the one still used today. He also used this model to explain the kinds of radiation atoms give off, and how that radiation can be used as a chemical fingerprint for detecting what elements are in a material.

PHYSICISTS

Are you up to date with the physics greats? Try this quiz to see how much you've learned.

Questions

1. How many groundbreaking papers did Albert Einstein publish in 1905?
2. What was the name of Isaac Newton's dog?
3. In which city was the first physicist, Thales, born?
4. What was Galileo's second work on heliocentrism, which earned him the Catholic Church's ire?
5. What dangerous elements did Marie Curie carry around with her?
6. What was Jocelyn Bell Burnell's research on?
7. What was Tycho Brahe's fake nose made of?
8. What did Albert Michelson and Edward Morley fail to find?
9. Who did Niels Bohr flee from?
10. At which university did Richard Feynman give his famous lectures?

Turn to page 216 for the answers.

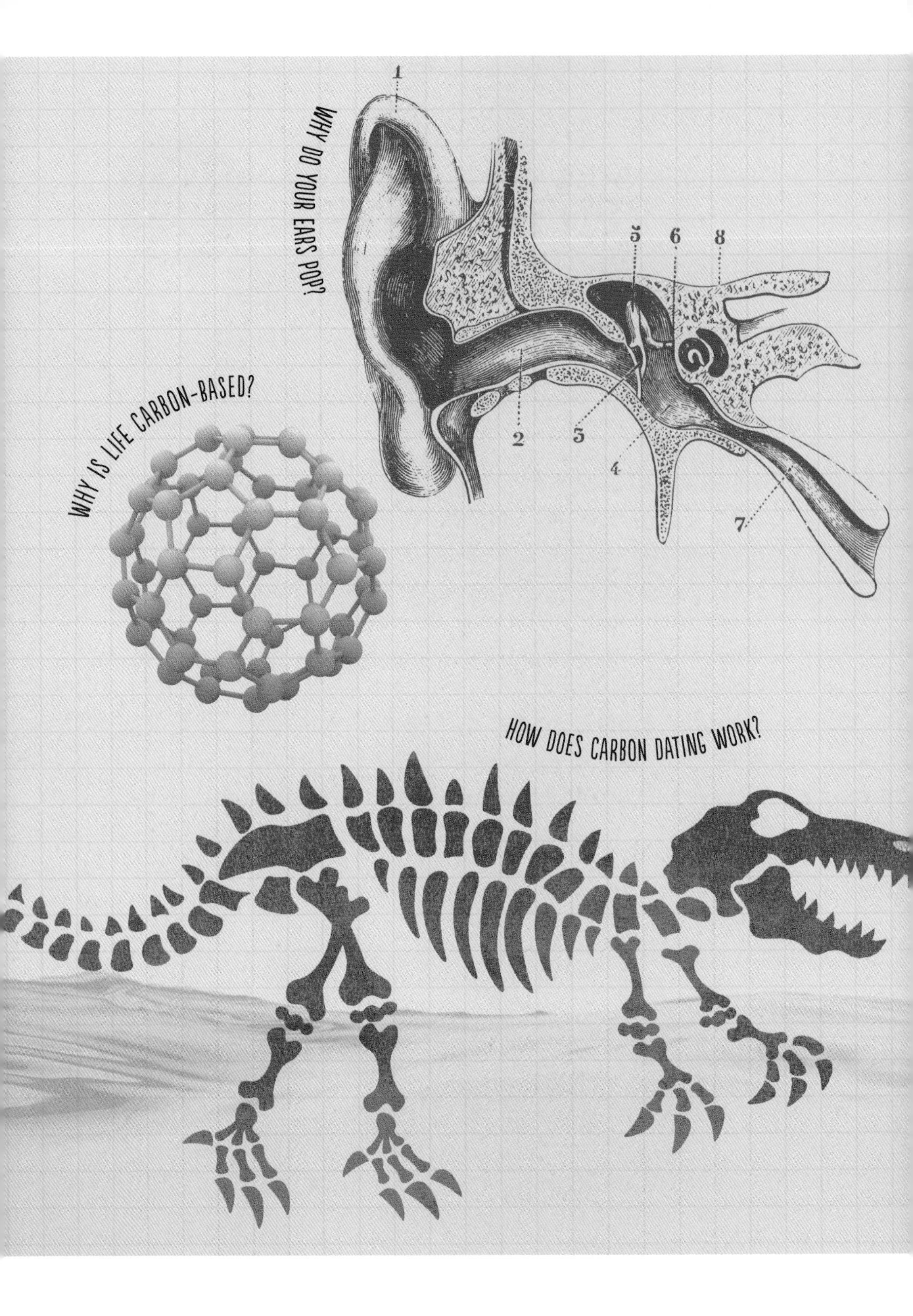
WHY DO YOUR EARS POP?
1
5
6
8
2
3
4
7
WHY IS LIFE CARBON-BASED?
HOW DOES CARBON DATING WORK?

HOW DO BACTERIA SWIM?

BIOPHYSICS

How do bacteria swim?

Bacteria are about as small as living things can get. They live in a world that is on such a different scale that it's hardly recognizable to us humans; the rules just aren't quite the same. So to deal with this, in order to get around bacteria swim by rotating their bodies with their motorlike flagella, or tails, rather than by kicking.

What a Drag

When you go swimming, you produce forward force by moving your arms and legs through the water, but bacteria are much smaller than us, so they feel more effect from the liquid around them. It would be like if you tried to swim through syrup. So instead of trying to push forward, bacteria have to use a different type of force: drag.

A Biological Motor

Within a bacterium is a type of biological circuitry. Rather than carrying blood (because a blood cell is about four times larger than the average bacterium) it carries electrons, like in an electrical circuit. This flow of electrons is then able to drive a motor in much the same process that we use to generate electricity (see page 106), but in reverse. This rotating motor causes the bacterium's coiled tail and its body to start rotating in opposite directions. This makes the bacterium twist through its viscous surroundings, essentially meaning that it pulls itself along using drag forces.

This is not a perfect process, however—the fluid moves randomly, meaning that bacteria still move all over the place, but the addition of this spinning movement means that they are able to create a preferred direction of movement. So while it may take some time, they will eventually get where they want to go.

Why is the shape of a red blood cell important?

Red blood cells play an incredibly important role in our bodies. They are carriers that take oxygen from our lungs to where it's needed. They are specially adapted to fit this purpose, right down to their shape. Red blood cells' distinctive shape allows for a faster transfer of gases through the body.

Biconcave Structure

The cells' shape is known as a "biconcave disk." This means that they are pushed in at the center on both sides—a bit like a ringed doughnut with only a thin film covering the central hole. This shape is useful because it gives the cell a greater surface area than a standard disk would have. (Surface area is how much exposed surface there is on an object.) The reason this is helpful is that the surface is where the cell is able to take in and give oxygen or carbon dioxide. So the greater the surface area, the more easily it's able to do this.

Surface Area

It's not just blood cells that have special shapes for increased surface area. Your lungs aren't just big sacks; they are full of alveoli, which are little air-filled hollow balls that bunch up like grapes. Having these alveoli increases the surface area of the lungs hundreds or even thousands of times, making the lungs much more efficient at taking in oxygen when we breathe.

How does carbon dating work?

You've probably seen a museum exhibit that shows a fossil or an ancient tree stump and a small plaque that dates it as millions of years old. Chances are that the given age was ascertained through radiometric dating. Carbon dating is one form of radiometric dating that uses the radioactive decay of carbon to analyze samples.

Carbon Dating

Every living thing contains a little bit of the element carbon-14. We take it in through the air that we breathe and it ends up in our bodies. Living organisms take in and use up carbon-14 at about the same rate, so there's always roughly the same amount in a body; however, when the organism dies it stops taking more in.

Carbon-14 is radioactive (see page 206). A radioactive substance decays over time; it also decays at a constant rate because of its half-life (see page 68). Scientists are able to use the half-life of carbon-14 to determine how long ago an organism died. They do this by comparing how much carbon-14 you would expect to see in a living version of the organism against how much there is in the sample you're analyzing. If there's half as much, then it died about 5,715 years ago; if there's a quarter, then it died about 11,430 years ago; an eighth means 17,145 years ago, and so on.

Beyond Carbon

Carbon dating is quite limited. Because of the relatively short half-life, it can only reliably be used to measure so far into the past. After about 50,000 years, or just under nine half-lives, the amount of carbon-14 becomes very small,

making it hard to detect. Fortunately, there are lots of other radioactive materials found in rocks and other substances that can be used to date objects such as fossils. The differing half-lives means that they can be used for lots of different purposes.

The longer a half-life is, the further back the dating method can be used (some, like thorium and samarium, have half-lives longer than the age of the universe), but the less precise it becomes. Carbon dating can get the date right to within about sixty years. Samarium dating, on the other hand,

FINDING FAKES

You might think that radiometrics would only be good for finding out fake fossils, but using half-lives to date objects is also used to find fake art! Between 1945 and 1963, there were a lot of nuclear tests across the world. One of the results of this is that many things made today have more radioactive material in them than those made before that time period. One of these things is a binding agent used to make paints. Any paintings made after 1963 have more radioactive material in them, and this is used to help identify fake paintings pretending to be from before that period.

has no upper limit in age but it only gives a date to within the nearest 20,000,000 years. Bioarchaeologists therefore have to use a mixture of different types of radiometric dating, depending on the approximate age of a sample, along with other techniques to try and determine a rough date for the creation of the object.

Why do your ears pop?

You've probably felt it before. Maybe on a plane or in a tunnel or while climbing a mountain. Suddenly your ears pop. It's an odd experience and you might wonder what it is and why it only happens in certain situations. Your ears popping is the equalization of pressure between your inner ear canal and your surroundings.

It's All About Pressure

The ear is a complex system with a lot of working parts. In the middle ear is a section called the Eustachian tube. It's a canal that connects your ears to the back of your nose. It is filled with air that is at the normal air pressure we experience every day on the ground, and is usually kept closed. When you're in a plane or a tunnel the air pressure surrounding you can change. Often it lowers and this means the air inside your inner ear is at a higher pressure than outside. This causes the air in the Eustachian tube to bulge and push against the rest of the ear, which can be uncomfortable and makes hearing harder.

Popping

For the pressure to equalize, the tube needs to open, allowing air to flow out. This happens very quickly and is the sensation we feel when our ears "pop." This can happen naturally, but it can also be induced through actions like yawning or swallowing. If the tube gets blocked by swelling or fluid when you're ill with a cold, the imbalance of pressure is what can cause dizziness and ringing in the ears.

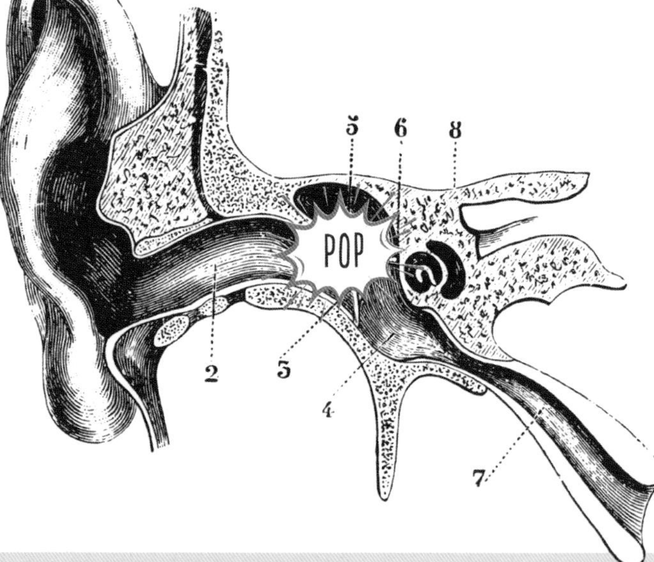

Why does the doctor leave the room when I have an X-ray?

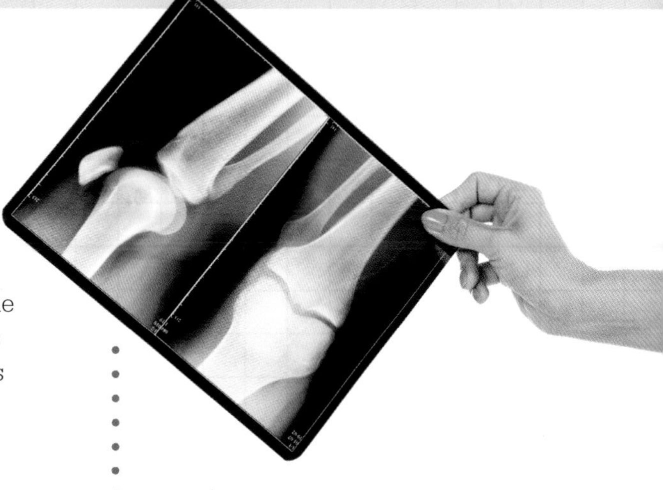

X-rays are a way of looking at the inside of your body without having to cut you open. They're incredibly useful and a staple of modern medicine. However, if you have one, you'll find that the doctor will leave the room. This may be a little alarming, but it's just because of radiation doses.

What Is an X-ray?

X-rays are a form of high-powered electromagnetic radiation (like light). Because of their high frequency, they are able to pass through things like skin and muscle but they are absorbed by our more dense bones. A type of reactive film is placed behind the person having an X-ray, and X-rays (the radiation) are fired at them briefly. Your bones are able to absorb the X-rays, but where there is no bone they pass through, hitting the film and causing it to change color. This gives doctors a picture of your insides, allowing them to check the bones for breaks and such without having to cut you open.

Exposure Levels

X-rays are a form of radiation that is perfectly safe in moderation. Each one is worth the equivalent of just a few days of normal background radiation from the environment. There is no effect on your health from having X-rays; even having upward of ten or twenty X-rays in a year won't harm you.

However, the dentist or doctor (often a radiographer) that gives the X-ray may repeat this procedure thousands of times a year. It is only when you could be exposed to the elevated radiation levels thousands of times for years on end that a normal X-ray scan could begin to pose a health risk. By leaving the room for the duration of the X-ray, doctors can limit their exposure across their career, thus keeping their exposure to a safe level.

Why are plants green?

The wondrous bounty of nature is all around us in the grasses, trees, and plants. There are a million varieties, but they almost all share a common element: green leaves. Plants are (probably) green because the Sun is green.

Green Sun

You might not think it, but the Sun is green. Like all stars, it gives off many different colors of light across the entire electromagnetic spectrum (even beyond visible), so the Sun is "white" as this is what you get when you mix all of the colors. However, the light that it gives off the most of is green, so we would call the Sun green. It doesn't look green to us, because before the light reaches our eyes it passes through the atmosphere, which scatters the light, making it seem yellow.

Photosynthesis

Leaves are green because chlorophyll is green. Chlorophyll is the part of leaves that absorbs sunlight in order to get the plant the energy it needs for photosynthesis (which is how a plant creates its food). So the question, then, is why chlorophyll is green. Some people think that leaves are green because that way they can absorb the Sun's abundant green light. However, if the leaves are green, then that means they absorb everything but green, because green is the only type of light able to bounce off and into your eyes. If they really were adapted to absorb as much light as possible, then they'd be black. While there are some black plants out there, they are few and far between. The truth is, it's not really known why most plants are green, but the leading theory is that plants get their fill of sunlight without the green light, and if they took it in it would just cause them to heat up, which could be damaging. By being green, they can safely reflect the green light away.

Why is water so important to life?

Water is needed by every living thing. From amoebas to elephants and everything in between, they've all got to have a drink. This is because water molecules are able to break down lots of different chemicals that living things need.

Chemical Scissors

Water molecules are made of two hydrogen atoms and one oxygen atom in a triangle shape. The oxygen atom is larger and has more electrons around it. This means that a water molecule is slightly more negatively charged at the tip and thus more positively charged at the back. This quality is called dipolar. This charged state on water molecules makes it the perfect chemical scissors. The charged tip and hydrogen atoms are able to provide a strong force that can be used to break apart other molecules. This means that water is able to dissolve lots of different chemicals into the base components that are needed for all sorts of processes in the body, such as breathing.

Other Uses

The ability of water to break down chemicals so blood can carry them around the body is incredibly important, but that's not its only purpose for living organisms. Water is central to many processes, from photosynthesis to respiration. Water is also able to help remove toxins from the body and, without it, any organism will dry up and no longer be able to take in the nutrients it needs, meaning that it dies. Water is therefore crucial to every single living organism. Water also forms a chemical bond called a "hydrogen bond," which is responsible for many of its useful properties. For example, hydrogen bonds cause lakes to freeze from the top down, creating a safe space for fish to live at the bottom during cold periods.

How radioactive is my food?

Radioactivity can seem scary, but it's both a great way of killing dangerous microbes to keep food safe and a perfectly natural part of our world. Even so, you might be concerned about eating irradiated stuff—but don't worry, your food is not very radioactive at all.

Measuring Radioactivity

Radioactivity is measured in a few ways. The standard unit of measurement is the sievert (Sv). It is a measurement of the amount of radioactivity absorbed by a living body and is the best way of measuring the risk to humans. Different parts of the body absorb radiation in different amounts, and it naturally reduces over time; just because you receive a 5-µSv (5-microsievert) dose from an X-ray doesn't mean it will stay in your body forever. There is a lot of radioactivity in the environment around us, so we receive a daily dose of about 1 µSv. A lethal dose of radiation is about 3,500 mSv (3,500 millisierverts, 3.5 million times greater than the daily dose).

Why Radioactive?

There are two main ways your food might become radioactive. The first is that it has been purposely irradiated by the food company. This is done in very small doses to sterilize foodstuffs. Administering a carefully controlled dose will kill microbes and other harmful things without damaging the food. Many other foods are naturally radioactive because of the chemicals they contain. Bananas are well known to be radioactive due to the potassium-40 within them, and consist of a dose of just 0.1 μSv, which is about 1 percent of normal daily exposure. You would need to eat 35,000,000 bananas in less than a few days to die of radiation poisoning from bananas. Other radioactive foods like potatoes, nuts, and beans contain similarly tiny doses of radioactivity.

Radioactive Bodies

To take in enough radiation to do damage to your body, you would either have to be standing near the site of a nuclear explosion or next to a nuclear power plant reactor. It can quickly prove fatal. One of the scariest things about radiation is that you can't see it, feel it, or smell it, though many people affected by radiation have reported being able to taste something strangely metallic.

If exposed to between 1 and 2 sieverts of radiation, a person would start to develop a mild headache and become very weak and tired. Though this would soon pass, in the long run their chances of getting cancers such as leukemia would increase significantly.

At between 3 and 4 sieverts, a person would start to feel very ill and become woozy. Their hair would start to fall out and their immune system would be compromised. Without significant treatment, they would die in less than a month.

Above 6 sieverts a person would become violently ill, have seizures, and vomit, until eventually within a few days they would die, even with significant medical help.

Because the effects of radiation sickness can be so deadly so quickly, any type of radioactive source is tightly controlled, and radiation levels in food are some of the most strictly regulated areas of manufacturing.

Why is life carbon-based?

Life here on Earth is all carbon-based. That means that the chemical that forms the basic building blocks on which life exists is carbon. It makes up nearly half of all the dry biomass (the combined weight of all living things) found on Earth and is the key component that the rest of the elements, like phosphorus and nitrogen, sticks to. This is because of carbon's specific properties.

Why Carbon?

Firstly, carbon is very abundant on Earth; there's enough of it around to form complex structures. Secondly, carbon is a very flexible element. It has four outer electrons that allow it to form four different bonds, which means it can combine into many different shapes (see pages 74 and 83). One of the more important things it can form is hydrocarbon chains. These are incredibly long molecules made of just carbon and hydrogen, which are perfect for creating complex molecules. Thanks to its multiple bonds, carbon is able to easily bond with other important chemicals such as oxygen, nitrogen, phosphorus, and sulfur. These chemicals are then able to bond onto the hydrocarbon chains that form proteins, which then make DNA, and so on.

Non-Carbon Life

There is no known reason why life can only be carbon-based. Astrobiologists study extreme creatures and examine exoplanets to allow them to speculate on what alien life there might be in the universe. They have long since hypothesized about silicon-based life-forms, as silicon shares the four-bond structure of carbon, but there could also be other bases. Some scientists have suggested that in different environments life might be able to form out of sulfur, boron, or even various metals that can create structures similar to those we see in organic life-forms. This is not to say, however, that alien life, were it to exist, has to be anything at all like what we see here on Earth—it could be completely different and beyond our wildest imaginings.

BIOPHYSICS

Got a grasp on the physics that makes you tick? Test out your biophysics knowledge with these questions.

Questions

1. How many average-size bacteria would you be able to fit in a blood cell?
2. What shape are red blood cells?
3. What is the name of the tube that connects your inner ear and nose?
4. How old is the oldest thing you could date using carbon-dating?
5. What type of radiation are X-rays?
6. What color is the Sun?
7. What shape are water molecules?
8. What is the most likely type of life-form apart from carbon-based?
9. How many bananas would you need to eat to die from radiation poisoning?
10. Which radioactive material would let you date something several times older than the universe?

Turn to page 216 for the answers.

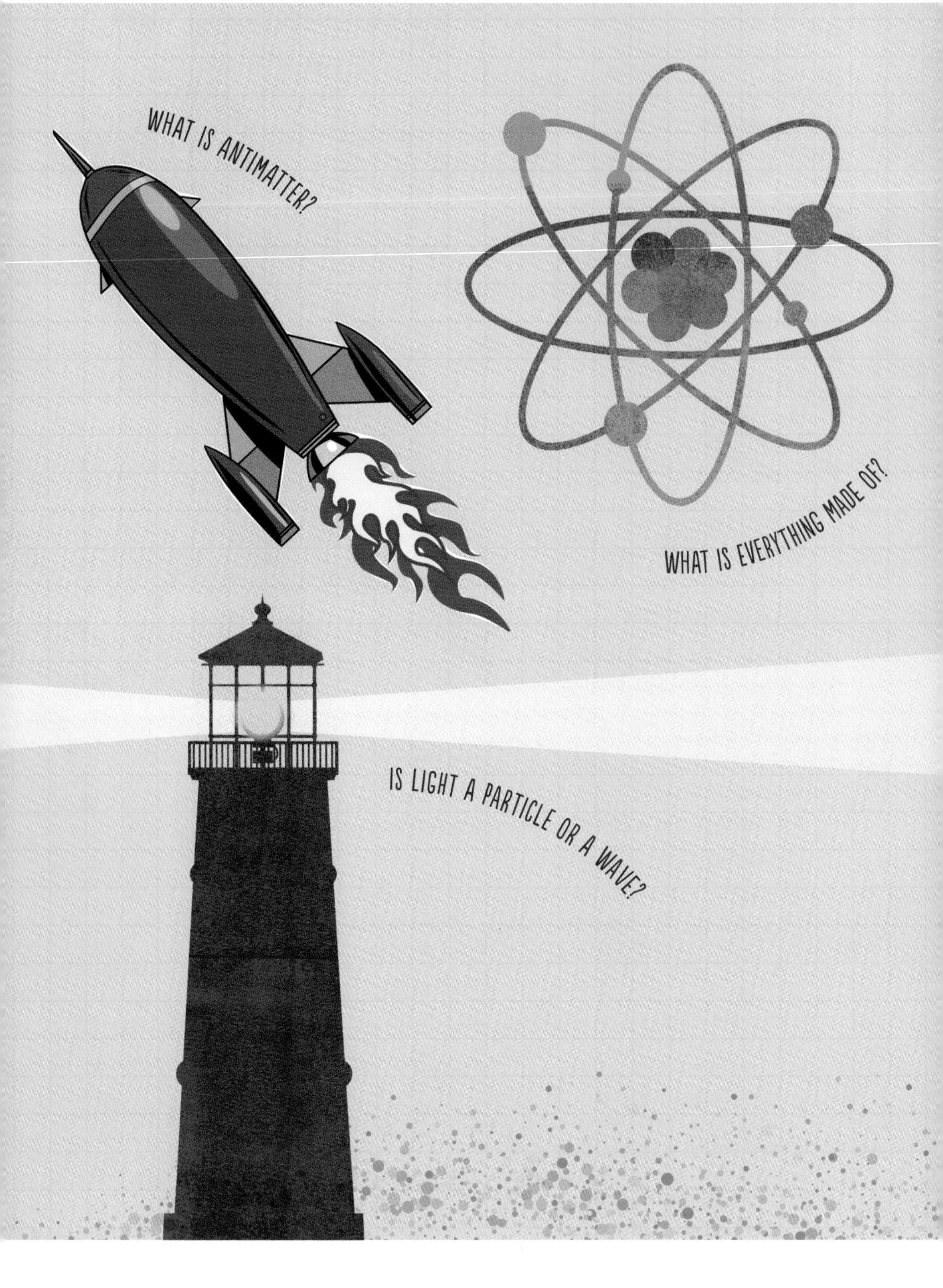

WHAT IS ANTIMATTER?

WHAT IS EVERYTHING MADE OF?

IS LIGHT A PARTICLE OR A WAVE?

PARTICLES

HOW LONG IS A PIECE OF STRING?

HOW WAS EVERYTHING MADE?

DOES THE PHILOSOPHER'S STONE EXIST?

What is everything made of?

Everything is made of atoms. From galaxies to giraffes and even you, on the most basic level, it's all constructed from atoms. But if everything is made from atoms, then what makes up these atoms themselves? Atoms are made of neutrons, protons, and electrons.

Inside the Atom

When we get down to the size of atoms, we need to start rethinking how we examine objects. You may have seen representations of atoms as clusters of different-colored balls, but this isn't quite right. The constituent parts of atoms don't really form a shape in the traditional way of thinking about it; instead, scientists tend to think about neutrons, protons, and electrons as a set of properties. Protons have a mass of 1 and a charge of +1. Neutrons also have a mass of 1 but no charge, and electrons have a mass of 0 (they do still have mass, but it's so small that it is mostly ignored) and a charge of -1.

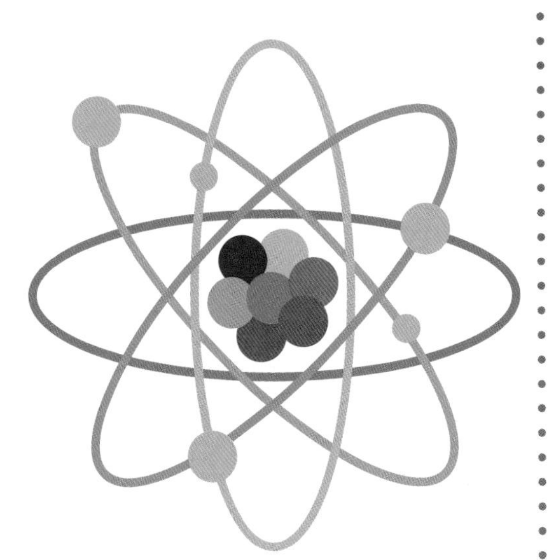

Atoms are made of a core called the nucleus, which is constructed of protons and neutrons. These are pulled together by the strong nuclear force, which is a duodecillion (that's 1 with 38 zeros after it) times stronger than gravity but works only on the very small scale of atoms. Protons have a positive (+1) charge and, like poles of magnets, will repel each other—so the neutrons help spread them out and act as buffers between the protons, meaning that they help to stabilize the nucleus. The protons and neutrons are similar in size, but the electrons are much smaller. Electrons have a negative charge, whereas all of the positively charged protons in the nucleus give it a positive charge. As in magnets, opposite charges attract, and so the electrons orbit around the nucleus.

Different Types of Atoms

An element is defined by the number of protons there are in the nucleus of a single atom of that element. The same element can, however, have different numbers of neutrons and electrons. These are known as isotopes and ions.

Isotopes are determined by the number of neutrons in the nucleus of an atom. All elements have multiple possible isotopes. The larger a nucleus is, the more different isotopes that element may likely have.

Ions are atoms that have more or fewer negatively charged electrons than positively charged protons, which means that the atom has an overall charge that depends on how many more or fewer electrons there are.

The number of protons, neutrons, and electrons inside an atom is what determines its properties, which in turn allows for the huge variety of stuff that we can see in the universe.

The Smallest Particle of All

Electrons are their own thing, but it turns out that protons and neutrons are made up of something else—quarks!

Quarks are not particularly well understood, but there are six different types: up, down, top, bottom, strange, and charm. Neutrons are made of two down quarks and one up quark, whereas protons are made of two up quarks and one down quark. Other types of quarks make up more exotic particles such as mesons. Quarks are best described by their properties. Up quarks have a two-third positive charge and down quarks a one-third negative charge, which accounts for why protons and neutrons have +1 and 0 charge respectively.

Quarks are also held together by the strong nuclear force, and it seems impossible to be able to get one on its own. When attempting to separate them, the energy needed to do it causes them to form two new quarks, making two pairs of quarks instead. It is not currently known if there is anything smaller than quarks.

Does the philosopher's stone exist?

Sought by alchemists throughout the ages—including such luminaries as Isaac Newton—the fabled philosopher's stone is an imagined substance that can turn base metals into gold, bringing with it enormous fortunes. In theory, with modern technology, it is possible to turn other elements into gold.

Making Gold

All elements are made out of atoms (see page 192), and an element is determined by the number of protons in its atoms. Gold has 79 protons in it, so all we need to do to turn another element into gold is either remove or add protons until it has exactly 79.

Removing protons from an atom is nearly impossible, and adding them isn't much easier. Adding a proton to a hydrogen atom (known as fusion) requires temperatures of millions of degrees and enormous pressures such as those found at the centers of stars. Even the (slightly) easier method of smashing large atoms into each other to add protons requires huge, cutting-edge facilities.

The biggest problem, though, is that the element with 78 protons, which would be the easiest and cheapest to convert into gold, is the even more valuable platinum. So although it's possible in theory, and has been done as a matter of proving the principle, no one would ever bother turning platinum into gold because it's far too expensive and time-consuming.

Giant Atoms

So what is this technology used for? Scientists use the ability to add protons to create and examine elements beyond those found naturally. While these huge atoms might only stay stable for a few fractions of a second, studying them can teach scientists a lot about how atoms themselves work. Oganessian is the atom with the most protons at 118. It was first made in 2002 in the Russian Joint Institute for Nuclear Research and was named after Yuri Oganessain, who played a leading role in its discovery.

Can you actually touch anything?

We are touching things all the time—you're touching this book right now! But what does that actually mean? Well, depending on how you define it, you might be able to touch everything or nothing at all.

Touching Nothing

The devil is in the details for this question, and it all comes down to what you define as touch. Traditionally, we might say that touch is when two objects have no space between them, such as a cup sitting on a table, but when we look at things in really close detail, it's not so simple.

We are constantly hovering—no matter if you're sitting in a chair or walking on the ground, there is always some space between you and the surface that's supporting you. If you were to zoom in close enough, you'd be able to see that the surface layers of the two objects are being held apart by the repulsive electromagnetic forces of the electrons, albeit in a solid and stable way. This space is present between everything, even inside the atom itself, so in this case nothing is ever touching.

Touching Everything

The idea that nothing ever touches is clearly a bit silly as clearly we can do it, so instead we might define touching as objects that are directly interacting with others. So when surface electrons are repelling those of another surface, then that is enough to call it touch. However, this leads to another issue: gravity and the electromagnetic force weaken over distance, which means that every atom in the universe is interacting in some way with every other atom. Many scientists stipulate that, for touch, the interaction must be of a certain strength and at a small enough distance, but these distinctions are arbitrary. So in a way we're always touching everything all the time.

Is light a particle or a wave?

The nature of light has been under examination since the tenth-century scholar Ibn Al-Haytham first discussed the idea in the *Book of Optics*. Yet it remains an area of much confusion and complexity, much of which centers around the question of whether light is a wave or a particle. It is, in fact, both.

Light as a Wave

You most likely think of light as a wave. It is, after all, often spoken about in terms of wavelength and it is part of the electromagnetic spectrum, much like radio waves. But it's not so simple. In 1672 Isaac Newton described light as small particles called "corpuscles," which became the leading theory. There were many problems with this approach, but it wasn't until Thomas Young's double-slit experiment of 1803 that waves took over as the dominant theory. Young's experiment involved a simple pair of slits and a back wall. In shining a light through the slits, it produced a pattern of light and dark fringes on the wall. This could only be a result of interference of the waves coming from the two slits. This seemed to prove conclusively that light was indeed a wave.

Light as a Particle

It took the genius Albert Einstein to throw a wrench in the works. In his day, the photoelectric effect was a well-known problem. By shining a light on a piece of metal, it would be possible to power a circuit as the light would knock the electrons out of the metal, letting them flow around the circuit. However, it was realized that the light always needed to be above a certain frequency. Einstein showed that this proves that light is a particle. If it were a wave, increasing the intensity or how long the light was on the metal would get the electrons flowing. As this wasn't the case, it meant that light was a particle whose energy was determined by its frequency, and it needed enough energy to break an electron out of the metal in one go.

Wave-Particle Duality

So there is proof that light is both a wave and a particle. The simple answer to this, then, is also the strangest: it's both. In fact, the double-slit experiment that Young used to show that light is a wave can also show that it is a particle. Furthermore, it's not just light that is both a wave and a particle. In 1924,

Louis-Victor de Broglie came up with a formula that showed all matter is able to act as a particle and a wave. Electrons (which were already well known to be particles) could be fired through the double slit, one electron at a time, and the fringe pattern seen by light waves would also appear. The conclusion from all of this is that all matter is able to exhibit both wave-like and particle-like properties simply depending on the circumstances. While this might not matter much on a day-to-day basis, when we examine light or other small particles, it becomes very important.

"The double-slit experiment that Young used to show that light is a wave can also show that it is a particle."

How can you look at tiny things?

Our world is filled with tiny things, but many of them are too small to ever see as they are smaller than light itself! So, how exactly are we able to know they're actually there and measure them? Tiny particles are not detected directly but rather by the effects they have in highly sophisticated experiments.

Cloud Chambers

Cloud chambers are sealed boxes that are then filled with some sort of vapor, such as water or alcohol. When a small particle passes through, it interacts with the vapor and causes a trail to form, in a similar way to the trails that planes leave behind. It is then possible to analyze these trails to learn about what made them.

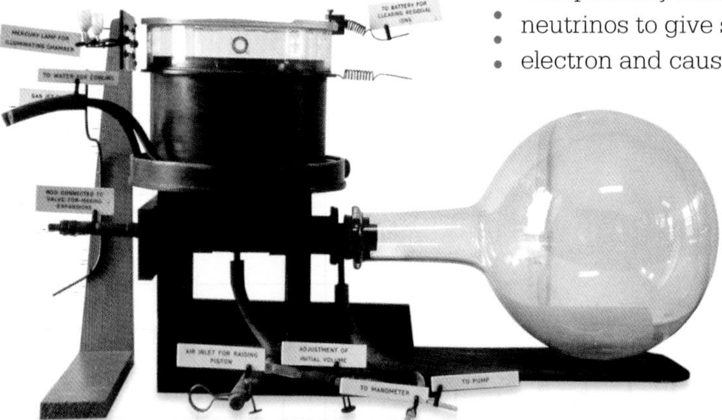

An electrical field is also often introduced into the cloud chamber, which affects the paths of any charged particles. This causes lightweight electrons to spiral in one direction, but in one experiment spirals moving in the other direction were found, leading to the discovery of the positron.

Neutrinos

Neutrinos are even smaller and less interactive than electrons (see page 208), so despite the fact that there are billions of them passing us every second, they are still incredibly difficult to detect. They are detected in facilities buried deep under the ground to keep them shielded from other types of cosmic radiation that might interfere. While the chances of it occurring are exceptionally small, it is possible for neutrinos to give some energy to an electron and cause it to accelerate rapidly, producing what's known as Cherenkov radiation. So, to detect neutrinos, enormous tanks are built, filled with electrons, and covered in highly sensitive cameras to detect the tiny amount of light the electron gives off.

Dark Matter Detection

Dark matter is a big unknown (see page 42), but one of the theories about it is that it's made of particles even smaller or less interactive than neutrinos! As it's not known exactly how these particles might work, there are a number of different types of detectors, all of them set up in deep mines to reduce interference from cosmic radiation. Some of these work in a similar way to neutrino detectors, but by using crystals at near absolute zero degrees, where a particle of dark matter may interact and cause a small amount of light to be generated. Others use a tank full of pure gas, which, if it interacted with dark matter, would produce a different element, which would float to the top and be collected. To date, however, no dark matter has been detected.

SMASHING PHYSICS

The Large Hadron Collider (LHC) at the CERN facility in Switzerland accelerates particles up to nearly the speed of light and then smashes them into each other. When it does this, the resulting explosion can create all sorts of strange particles. These particles then either interact or decay into other things, which can then be picked up by the surrounding ATLAS detector. Using this, the LHC has found some incredible things such as the Higgs boson, which is the particle through which gravity works.

How was everything made?

We're not quite sure what happened in the beginning. But for a few fractions of a second there was suddenly a lot of energy in a very small space. From this short time came everything in today's universe. But while it was all the same stuff then, it's now all sorts of different stuff made out of the various elements of the periodic table. Everything is made of elements and almost all elements are made by stars.

The First Things

At about one second after the big bang, the universe had cooled enough that protons and neutrons were able to form out of the intense energy. For the next 20 minutes or so, the protons and neutrons were able to fuse, forming lots of hydrogen, helium, and just a tiny amount of lithium. And then for about 150 million years, nothing much happened. Over this time period, the free-floating hydrogen and helium particles began to clump together into enormous gas clouds, which eventually collapsed in on themselves under gravitational forces and formed the first stars.

Solar Furnaces

Stars are fueled by the fusing of hydrogen atoms to make helium in their core. So for the first few hundred million years, the universe consisted only of these furnaces converting hydrogen into helium. However, as the first large stars started to die, something changed. As the stars ran out of fuel, they began to collapse inward again, and this caused them to get even hotter than before, letting them fuse together the remaining hydrogen and helium into new elements such as oxygen, nitrogen, and carbon. If the star was large enough,

this cycle of collapsing, heating up, and then fusing new elements could keep going on, creating the first twenty-six elements all the way up to iron.

Beyond Iron

When the core of a star reaches iron, the process of fusing the heavy elements now needs more energy than it creates, so stars stop producing new elements and collapse down into steadily cooling balls of matter. However, if a star is really big when it collapses for the last time, the overwhelming pressure causes its core to undergo a runaway thermonuclear reaction that causes an enormous explosion: a supernova. The blast from this supernova is so enormously powerful that it is able to fuse large amounts of elements up to and beyond iron. A supernova also jettisons these huge amounts of elements out into the universe.

How the Elements Make It to Earth

The materials that come from collapsing stars and supernovas are able to float out into the universe and begin the process all over again, with the new elements forming parts of gas clouds, which collapse and are able to make new stars and planets. The Sun is a star made of the remnants of this process occurring twice before, which means that its planets (which are made from the same stuff as the star they orbit) can be very rich in elements. Despite the universe being made of 75 percent hydrogen, it accounts for a much smaller percentage of the Earth because the stuff it's made from has already been through two stars.

What is antimatter?

There is a strange dark side to all of the possible particles that make up our world. They all have mirrored versions of themselves known as antiparticles.

A Mirror Image

Antimatter is made of the same stuff as normal matter but with one difference: The properties are switched. A positron is the antimatter version of an electron. It's the same size and mass, and acts exactly the same, except that its charge is +1 rather than an electron's -1. Antiprotons and antineutrons are made of antiquarks, and if they are matched up with positrons it is possible to create anti-copies of the elements of the periodic table. How exactly these antimatter elements would work, and whether they would create the same things as matter does, or an entirely different universe, is not known.

Never the Two Shall Meet

Matter and antimatter do not get along—when they get together, things get explosive. If an antimatter particle comes into contact with its opposing matter particle, the two of them annihilate. They are both instantly turned from particles into a huge amount of pure energy. It's because of this that antimatter doesn't stick around in our universe for very long. Almost as soon as any is created through particle collisions or other interactions, they annihilate with some matter back into energy.

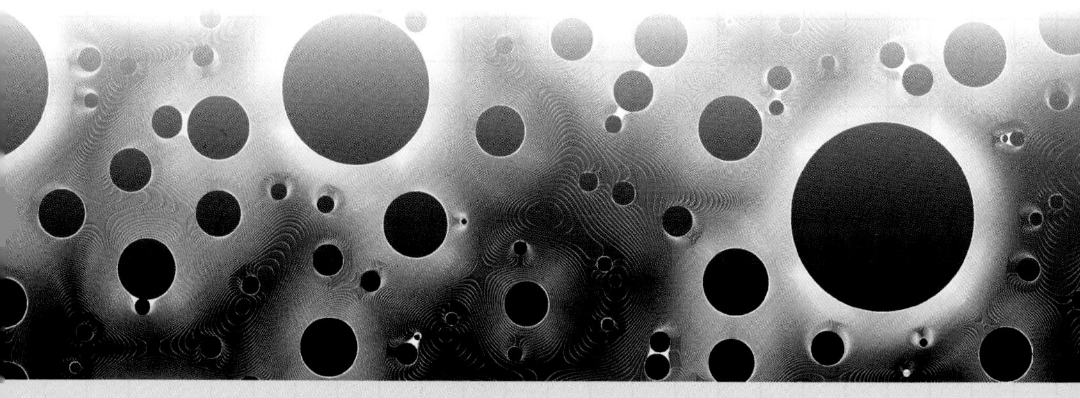

Why Matter?

The big question is: why is our universe full of matter and not antimatter? Why is it that there seems to have been at the beginning of the universe more matter when we might expect there to be the same amount of antimatter as matter? In short, nobody is really sure. It's possible that antimatter is for some reason less stable, or perhaps there are pockets of matter and antimatter in the universe and we just happen to be living in a matter area. For now, there is no real answer, but scientists continue to look.

MAKING ANTIMATTER

In science fiction, it is used as fuel for spaceships and the warheads of bombs for intergalactic emperors, but antimatter can also be created in the real world. Many high-energy experiments in labs across the world create antiparticles for study, and positron emission tomography (PET) scanners in hospitals use generated positrons to be able to take detailed images of the human body. Scientists have even been able to create antihydrogen, consisting of a single positron orbiting around an antiproton. The antihydrogen atoms existed for about a quarter of an hour before they annihilated. Research is still in its infancy, but in the future it is thought that because of the huge amount of energy it contains, it could take only a single gram of antimatter to fuel a spaceship all the way to Mars!

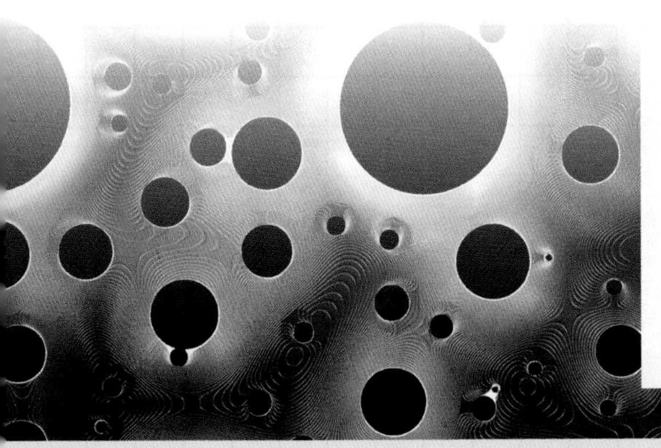

How long is a piece of string?

The traditional answer to this age-old question is: "as long as you cut it." But once you cut your piece of string, you still need to measure it. Measuring is not so simple when you get down to a really small scale—you can never be sure how long things are.

The Problem of Measuring

Think of a normal 12-inch wooden ruler. Along its length are black lines that indicate each fraction of an inch. If you were to take a piece of string, you could place it next to this ruler and measure it out. You might measure your string to be 5 inches long, but that's not going to be exactly correct. Your string probably ends a little bit before or after the marking. If your ruler is good enough it might mark on every eighth of an inch as well to give you 5.125 inches, or maybe even every sixteenth of an inch, giving you a more precise measurement at 5.1875 inches. Even then, if you were to look through a microscope, you would see that the string still doesn't stop on the marking. This problem will happen no matter what equipment you use to measure.

Uncertainty

The amount by which we are inexact is called uncertainty. Uncertainty not only occurs when measuring length but also time, weight, and literally everything else. You might think of this as just a problem with how we measure things, but it seems that it might be a more fundamental part of the universe. When you get down to the very small things, on the scale of individual particles, quantum effects start to make things a bit weird.

The Heisenberg Uncertainty Principle (published 1927 by Werner Heisenberg, pictured right) states that the amount of uncertainty about the position of a particle multiplied by the amount of uncertainty about its speed must always be above a given value.
What this means practically is that when you try to make more precise measurements of where something is, it will start to move more unpredictably, and when you try to measure its speed, you will have less idea of where it is.

Applying the Principle

The Heisenberg Uncertainty Principle means that we can never really know where electrons are around a nucleus, or the true speed of relativistic particles (those traveling at near the speed of light). It's not just speed and position, though; the principle also applies to energy and time, which can mean that particles can pop into existence from nothing so long as it's only for a short amount of time. In everyday life, the Heisenberg Uncertainty Principle doesn't much matter. The value that uncertainties must be above for the principle to come into effect is incredibly small. So when you try to measure your piece of string, you may not ever know perfectly where the end is and how long it is, but you can probably make a good enough measurement anyway.

> "When you try to measure your piece of string, you may not ever know perfectly where the end is."

What is radioactivity?

Radioactivity is something you have heard of, probably as something very dangerous and to be feared in the context of radioactive bombs, radioactive waste, and nuclear power plant disasters such as Chernobyl. But radioactivity doesn't come from some advanced technology; it comes from the decay of unstable atoms.

Types of Decay

There are three types of radioactive decay, each named after one of the first three letters of the Greek alphabet and made of different things.

Alpha decay: The release of a bundle of two protons and two neutrons (a helium nucleus).

Beta decay: The release of an electron and neutrino.

Gamma decay: The release of a high-energy electromagnetic wave.

The different types of radiation come from different processes and have different properties, so depending on the type of material being handled, they need to be treated differently. One radioactive source can produce multiple types of radioactivity.

Alpha

Some atoms get so large that it is difficult for them to be able to stay together. In order to try and reduce its size and become more stable, the atomic nucleus might eject two of its protons and neutrons as an alpha particle, which will cause the atom to change into a different element and become more stable. This process can happen more than once to a single atom. Alpha particles can be fairly dangerous. Their large size means they

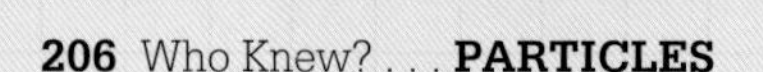

can easily damage cells in the human body, causing things like cancer; however, alpha particles also can't travel very far and are easily contained by something as thin as a sheet of paper. This means that alpha sources are only really dangerous in large quantities or if ingested.

Beta

In beta decay, in order to try and become more stable, a neutron in the nucleus of an atom changes into a proton. This process also creates an electron that is jettisoned from the nucleus at high speed as a beta particle. It is also possible for a proton to change into a neutron, which will release a positron instead. Beta particles are less able to damage people, as they are much smaller than alpha particles, but they are also able to travel much farther and are more difficult to contain—you would need something like aluminum to properly shield yourself from their radioactivity.

Gamma

Some atomic nuclei have too much excess energy in them, and they get rid of it by firing it out as a high-energy electromagnetic wave. Gamma waves are like any other electromagnetic wave such as radio or light, but they are far more intense and sit at the top of the electromagnetic spectrum. Gamma rays are able to penetrate skin easily and are the most dangerous type of radiation as they can travel very far. For this reason, gamma sources are often kept in thick lead boxes.

What is a neutrino?

There are literally trillions of neutrinos passing through your body every single second, but you never notice them. Neutrinos are tiny, nearly interactionless particles produced by atomic reactions.

Smaller Than Electrons

Neutrinos are a bit like electrons, only they're completely chargeless, much smaller, and have at least three million times less mass than an electron. Neutrinos are so small and (physically speaking) uninteresting that they hardly interact with anything, which is why they are so difficult to detect despite being around in such vast quantities. They are so uninteractive that they weren't first detected until 1956, and even today detecting them is a very difficult task that requires highly specialized equipment (see page 198). There are three different types—electron neutrinos, muon neutrinos, and tau neutrinos, though a given neutrino seems to be able to swap between the three states randomly.

Where Neutrinos Come From

Neutrinos are produced in a number of different ways. Radioactive decay (see page 206) is able to produce them, as are nuclear reactors and bombs, and they can also be created in particle accelerators. They are also very common in space, being created by supernovas, neutron stars, and, crucially, stars. Our own Sun produces neutrinos as a by-product of the fusion processes that occur in its core. It produces neutrinos in such enormous quantities that here on Earth millions of miles away, every square inch of the Earth receives over 500,000,000,000 passing through it each second.

PARTICLES

Able to separate your neutrons from your neutrinos? Test your knowledge with this quiz.

Questions

1. What is the name for the particles that are made up of protons and neutrons?
2. What's the easiest element to turn into gold?
3. What force keeps everything from touching?
4. Is light a particle or a wave?
5. What is smaller, a neutrino or an electron?
6. What device is used for looking at electrons?
7. What fuel do stars fuse together to make the element helium?
8. What is the antiparticle of the electron?
9. What principle makes knowing where something is difficult?
10. What are the three types of radioactivity?

Turn to page 217 for the answers.

WHAT MAKES STARS TWINKLE?

HOW FAST IS GRAVITY?

WHAT MAKES ATOMIC BOMBS EXPLODE?

WHY DOES HEAT RISE?

WHAT ARE COMETS MADE OF?

HOW DO MAGNETS WORK?

QUIZ ANSWERS

QUIZ ANSWERS

Celestial Bodies Quiz Answers

1. Distorts it
2. The Moon
3. Above 370 miles
4. 10,000°F
5. Halley's Comet
6. Meteors
7. Kuiper Belt
8. The Southern Cross
9. Six hours
10. 2492

Cosmology Quiz Answers

1. Pizza
2. The event horizon
3. The core of a supernova
4. 2.7 degrees kelvin
5. Nearly 70 percent
6. 13.8 billion years old
7. Four
8. The big crunch
9. Cosmic microwave background
10. The Chandrasekhar Limit

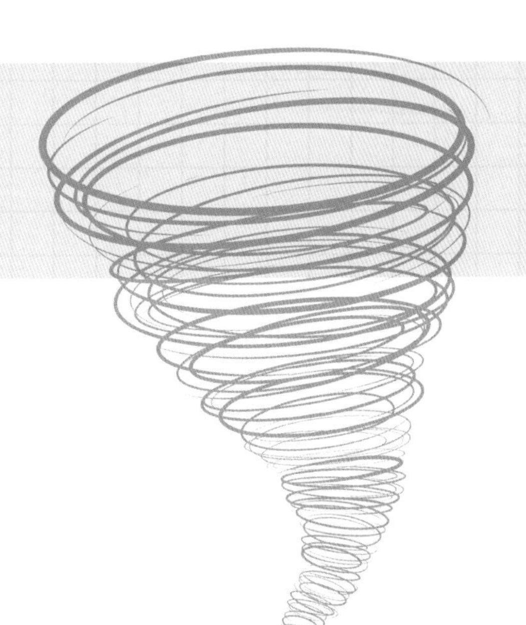

Weather Quiz Answers

1. Between 12°F and 32°F
2. Water droplets
3. Counterclockwise
4. 30–160 feet
5. About 1.8 miles
6. Nitrogen
7. Hydrogen and oxygen
8. Chaos theory
9. The effect where hurricanes push seawater onto land
10. The aurora australis

Materials Quiz Answers

1. Tantalum hafnium carbide
2. Its ability to make metal bonds
3. About 30 years
4. Carbon
5. Static
6. -328°F
7. Vibration
8. Plastic polymers
9. They are used in MRI machines
10. In the Earth's crust

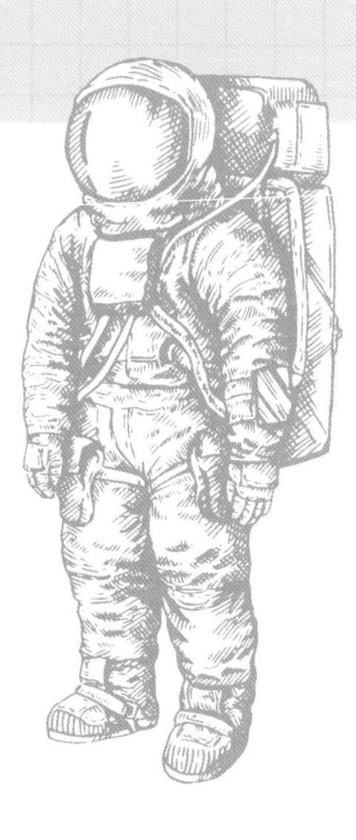

Technology Quiz Answers

1. Chocolate
2. Silicon dioxide
3. -455°F
4. Pressure
5. Graphite
6. Fusion and fission
7. The bottom
8. Water
9. LED (light-emitting diode)
10. Polarization

Computers and Electronics Quiz Answers

1. Heat-assisted magnetic recording (HAMR)
2. Radio waves
3. Over 400 amps
4. Electromagnets
5. Light
6. The voltage is decreased
7. Solar power
8. Qubits
9. Lithium-ion and nickel-cadmium
10. Three: red, blue, and green

Fundamental Physics Quiz Answers

1. You get heavier
2. Psi-function
3. Energy
4. Blue-shift
5. 3 Sigma
6. Platinum-iridium
7. One
8. Permanent magnets and electromagnets
9. Relativity
10. Entropy

Forces Quiz Answers

1. Buoyancy
2. 89 percent
3. Torque
4. Terminal velocity
5. Friction
6. A black hole
7. Five
8. 160°F
9. Solids
10. A non-Newtonian fluid

Physicists Quiz Answers

1. Four
2. Diamond
3. Miletus (modern-day Turkey)
4. Dialogue
5. Polonium and radium
6. Pulsars
7. Brass
8. Luminiferous ether
9. The Nazis
10. California Institute of Technology (Caltech)

Biophysics Quiz Answers

1. Four
2. Biconcave
3. The Eustachian tube
4. 50,000 years old
5. Electromagnetic
6. Green (or white)
7. Triangular
8. Silicon-based
9. 35,000,000
10. Samarium

Particles Quiz Answers

1. Quarks
2. Platinum
3. Electromagnetic force
4. Both
5. A neutrino
6. A cloud chamber
7. Hydrogen
8. A positron
9. The Heisenberg Uncertainty Principle
10. Alpha, beta, and gamma

INDEX

CREDITS

12 © kdshutterman | Shutterstock • 13 © Sakinakhanim | Shutterstock, © ClickHere | Shutterstock • 14 © Vladi333 | Shutterstock • 15 © DreamStockIcons | Shutterstock, © Jason Winter | Shutterstock • 16 © helgascandinavus | Shutterstock, © popicon | Shutterstock • 17 © Mary Frost | Shutterstock • 18 © VectorPot | Shutterstock • 21 © d1sk | Shutterstock , © Michael Rosskothen | Shutterstock • 22, 37 © NASA images | Shutterstock • 23 © PlanilAstro | Shutterstock, © ArtMari | Shutterstock • 24 © Orkidia | Shutterstock • 25 © Artur Balytskyi | Shutterstock • 26 © Nicku | Shutterstock • 28 © WWWoronin | Shutterstock • 29 © Hoika Mikhail | Shutterstock • 30 © GoodStudio | Shutterstock • 31 © zuper_electracat | Shutterstock • 32 © Somchai Som | Shutterstock • 36 © ASAG Studio | Shutterstock • 39 © Dotted Yeti | Shutterstock •40 © Tomas Ragina | Shutterstock, © BullsStock | Shutterstock • 41 © Yurkoman | Shutterstock • 42 © sakkmesterke | Shutterstock • 44 © anigoweb | Shutterstock • 45 © TheFarAwayKingdom | Shutterstock • 46 © CatbirdHill | Shutterstock • 48 © Martin Capek | Shutterstock • 49 © Pyty | Shutterstock • 50 © Artur Balytskyi | Shutterstock • 54 © kelttt | Shutterstock • 55 © Reinekke | Shutterstock • 56 © Pogorelova Olga | Shutterstock • 58 © Potapov Alexander | Shutterstock, © Vector FX | Shutterstock • 59 © TAW4 | Shutterstock • 60 © Sylvie Corriveau | Shutterstock • 61 © Yuravector | Shutterstock, © Webicon | Shutterstock • 62 © suns07butterfly | Shutterstock •66, 90, 99, 116, 166 © Everett Collection | Shutterstock • 67 © RioAbajoRio | Shutterstock • 68 © Daniel Prudek | Shutterstock • 69 © solomon7 | Sutterstock, © janista | Shutterstock • 70 © Mr.adisorn khiaopo | Shutterstock • 71 © Leremy | Shutterstock, © Bettmann | Getty Images • 72 © cyo bo | Shutterstock • 74 © Manutsawee Buapet | Shutterstock • 75 © vitdes | Shutterstock • 76 © Pixel Embargo | Shutterstock • 80 © TerryM | Shutterstock • 81 © dharmastudio | Shutterstock • 82 © Dn Br | Shutterstock • 83 © Maisei Raman | Shutterstock, © AlexVector | Shutterstock • 84 © RedlineVector | Shutterstock • 85 © Babich Alexander | Shutterstock • 86 © kirill_ makarov | Shutterstock • 87 © Bettmann| Getty Images • 88 © eurovector | Shutterstock, © Alberto Zornetta | Shutterstock • 89 © Kyryloff | Shutterstock • 90 © BullsStock | Shutterstock • 91 © alex74 | Shutterstock • 92 © Alex Staroseltsev | Shutterstock • 96 © Ink Drop | Shutterstock • 97 © MMvector | Shutterstock • 98 © mkos83 | Shutterstock • 100 © Tuomas Lehtinen | Shutterstock • 101 © Francois Poirier | Shutterstock • 102 © Stocksnapper | Shutterstock • 103 © vinap | Shutterstock • 104 © HomeArt | Shutterstock • 105 © Champ Nattapon | Shutterstock • 106 © Makhnach_S | Shutterstock • 107 © Brovko Serhii | Shutterstock • 108 © Martial Red | Shutterstock • 109 © Amin Van | Shutterstock • 110 © Yuravector | Shutterstock • 111 © dwersteg | Shutterstock • 112 © Maksym Drozd | Shutterstock • 117 © Ficus777 | Shutterstock • 119 © Nikolai Tsvetkov | Shutterstock • 120 © jijomathaidesigners | Shutterstock • 122 © TRONIN ANDREI | Shutterstock • 123 © Outer Space | Shutterstock, © Science History Images | Alamy • 124 © AlexLMX | Shutterstock • 126 © AnastasiaSonne | Shutterstock • 127 © pukach | Shutterstock • 128 © ShutterStockStudio | Shutterstock • 130 © shurkin_son | Shutterstock • 131 © viewgene | Shutterstock • 132 © Chones | Shutterstock • 134 © Sharpner | Shutterstock • 135 © Esenin Studio | Shutterstock • 136 © Maisei Raman | Shutterstock • 140 © NASA | Dembinsky Photo Associates | Alamy • 141 © gjebic nicolae | Shutterstock, © ClickHere | Shutterstock • 142 © Bodor Tivadar | Shutterstock • 143 © Mike Flippo | Shutterstock • 144 © leedsn | Shutterstock, © Spiroview Inc | Shutterstock • 145 © Eric Isselee | Shutterstock • 146 © Vertyr | Shutterstock • 148 © ArtHeart | Shutterstock • 149 © AkimD | Shutterstock • 150 © Daniel Prudek |Shutterstock • 151 © studiostoks | Shutterstock • 152 © Pavel Chagochkin | Shutterstock • 153 © grynold | Shutterstock • 154 © stockphoto-graf | Shutterstock, © Bodor Tivadar | Shutterstock • 158 © Alpha Historica | Alamy, © Gado Images | Alamy • 159 © Dr Project | Shutterstock, © dramaj | Shutterstock • 160 © iryna1 | Shutterstock • 161 © Utekhina Anna | Shutterstock, © VikaSuh | Shutterstock • 162 © Georgios Kollidas | Shutterstock • 163 © Glasshouse Images | Alamy • 164 © Print Collector | Getty Images • 165 © Prachaya Roekdeethaweesab | Shutterstock, © CoolVectorStock | Shutterstock • 167 © eAlisa | Shutterstock • 168 © Daily Herald Archive | Getty Images • 169 © Jurik Peter | Shutterstock • 170 © Time Life Pictures | Getty Images • 171 © Emilio Segre Visual Archives | American Institute of Physics | Science Photo Library | • 172 © Everett Collection Inc | Alamy • 173 © elnavegante | Shutterstock • 174 © Arkadivna | Shutterstock • 178 © Macrovector | Shutterstock • 179 © Alex Mit | Shutterstock, © crystal light | Shutterstock • 180 © matsabe | Shutterstock • 181 © Jag_cz | Shutterstock • 182 © Morphart Creation | Shutterstock • 183 © Dmitry Lobanov | Shutterstock • 184 © tanatat | Shutterstock • 185 © SiiKA Photo | Shutterstock • 186 © SeDmi | Shutterstock • 187 © Kaiskynet Studio | Shutterstock • 188 © ynse | Shutterstock • 192 © Jemastock | Shutterstock, © jijomathaidesigners | Shutterstock • 193 © Ezume Images | Shutterstock • 194 © Kumer Oksana | Shutterstock • 195 © Dima Sobko | Shutterstock • 197 © astudio | Shutterstock • 198 © The Print Collector | Alamy • 199 © general-fmv | Shutterstock • 200 © Natali art collections | Shutterstock • 202 © GiroScience | Shutterstock • 203 © studiostoks | Shutterstock • 204 © Fedorov Oleksiy | Shutterstock, © xpixel | Shutterstock • 205 © Alpha Historical| Alamy • 206 © KREML | Shutterstock • 207 © BlueRingMedia | Shutterstock • 208 © Fermi National Accelerator Laboratory | Science Photo Library •